The story of
DINOSAUR
soft tissue

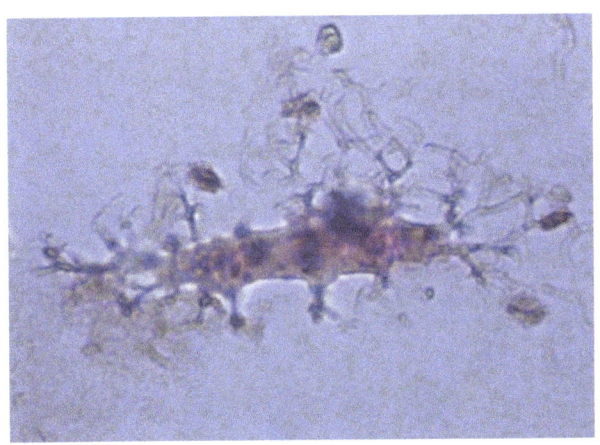

osteocyte found in Nanotyrannus   *Mark Armitage*

by Ellen J. McHenry

ISBN 979-8-9939064-1-6

Printed in the U.S.A.

Retailers can order directly from IngramContent.com.
Ingram has no connection to any facts or opinions stated in this book.
Other inquiries: ejm.basementworkshop@gmail.com

5.15.26

Some of the dinosaurs and reptiles mentioned in this book:

Tyrannosaurus rex

Edmontosaurus

Cacops

Dimetrodon

Apatosaurus

CHAPTER 1: Early discoveries

This story has a beginning, but it doesn't have an end. It is an on-going story that is still happening. Even while you are reading this, scientists are busy doing experiments and making new discoveries about what's inside dinosaur bones. This book will probably have to be updated in a few years when the new discoveries are published and made public.

In order to understand some of the key ideas in this story, you'll have to learn a little bit about anatomy (the structure of animal bodies) and about cells (the individual units that fit together to make tissues such as skin, muscles, bones, and organs). Much of the science we'll discuss is the same for people, so, as an added bonus, you'll be learning about your own body, not just dinosaur bodies. Some parts of this book might read more like a biology or physics text rather than a story, but hopefully those science lessons will be very interesting, and they will definitely help you to follow the story about dinosaur bones with greater understanding.

Our story starts in the year 1954. To those living in the 1950s, the world seemed very modern. They had telephones, radios, television, cars,

and airplanes. DNA had been discovered in 1953. The world's first nuclear power plant began generating electricity in 1954. In the world of biology, microscopes had been around for over two centuries and by now they were very good.

All of the major types of body cells had been discovered even if they did not yet know the function of all the little parts (organelles) inside them. Chemists had discovered all of the naturally occurring elements on the Periodic Table (everything up to uranium, number 92) and were working on extending the table as far past 92 as they could by creating super-heavy man-made elements. Chemists were becoming very good at extracting and analyzing all kinds of new molecules, both organic and inorganic. They were making use of new technology such as x-ray diffraction and radioactive isotopes. You don't need to understand these techniques,

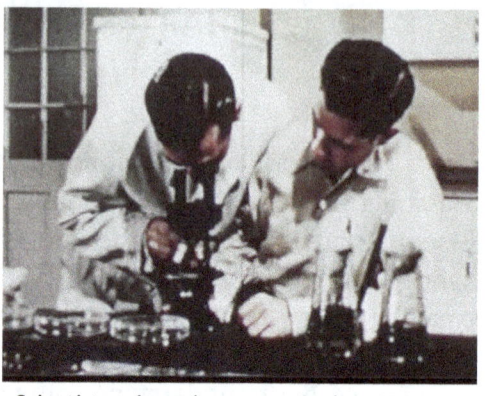

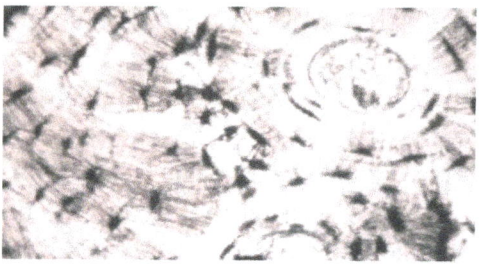

Scientists using microscopes in the 1950s to look at bone cells, shown below.

you just need to know that while ordinary people in 1954 were sitting in front of small black and white television screens with poor picture quality, scientists in labs were making quite advanced discoveries.

In 1954, the Carnegie Institution of Washington D.C. published their annual year book in which they reported on all the work that their researchers had done during the previous year. There were reports from the departments of astronomy, geophysics, plant biology, embryology, genetics, and archaeology. One of the articles from the geophysical department was entitled, "Organic constituents of fossils." In this article, the researcher reported on some of the molecules he found while examining fossils of several types of dinosaurs including a stegosaurus and a mosasaurus. Because of the assumed ages of these specimens (100 to 150 million years) he was surprised to find that they still contained tiny pieces of protein called **amino acids**. What are amino acids? It's time for our first biochemistry lesson.

2

WHAT ARE AMINO ACIDS?

Amino acids are the "building blocks" that cells use to make proteins. You are already familiar with the word "protein" because you have probably been told to eat plenty of protein (fish, meat, eggs, beans, seeds, nuts) as part of a healthy diet. Our bodies digest the proteins we eat, breaking them down into their smallest units, amino acids. (This is like tearing apart a Lego® structure until it is nothing but a pile of individual bricks.) Our bodies then use the tiny amino acids to build their own protein molecules. Some proteins are part of large structures such as muscles and skin. Others have interesting jobs such as carrying oxygen (hemoglobin), fighting germs (antibodies), forming microscopic "roads" inside our cells (microtubules), regulating our blood sugar levels (insulin), or carrying messages between cells (hormones and cytokines).

Amino acids are made of atoms. This diagram shows the basic structure of all amino acids. The letters represent different types of atoms. **C** is carbon, **H** is hydrogen, **O** is oxygen, and **N** is nitrogen. It's the COOH part (highlighted in yellow) that classifies this molecule as an acid.

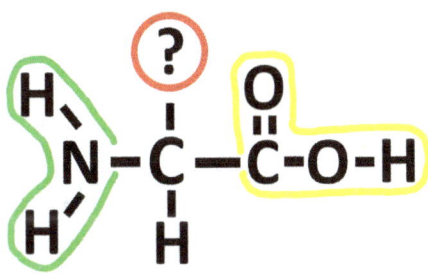

However, amino acids don't taste sour like lemons do. The question mark shows the place where different groups of atoms can be attached to this basic structure. Each amino acid has a unique grouping of atoms attached here.

The smallest and simplest amino acid is glycine, which has only a hydrogen atom, H, in place of the question mark. Coming in second place for the smallest is alanine, with only a C and three H's at the question mark. As we will soon see, aminos can attach end to end to make a long string called a peptide. But this is enough information for now.

GLYCINE

ALANINE

3

How did the researcher isolate and identify such tiny molecules in a piece of rock (a fossil)? First, he dissolved the fossils in very strong acids. Then this solution was passed through a special filter that got rid of all the minerals. Then the solution was dropped onto chromatography paper. An actual image from his 1954 report is shown here. ALA is alanine, and GLY is glycine. It just looks like smudges on a paper to us, but

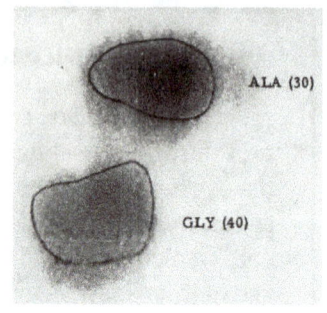

the scientist knew what he was looking at. He also used some radioactive isotopes (atoms that are unstable and falling apart) to help him identify what he was looking at. Radioactive atoms are easy to detect, so they are often used in analysis.

The researcher detected 6 different amino acids in all his fossil samples. Since there are only 20 amino acids that make up all proteins, finding 6 was quite significant. Surprisingly, the oldest fossil contained the highest amount of amino acids. How long can these little organic molecules last before they fall apart? Could they really last millions of years?

In 1963, a group of researchers in Oxford, England, did a very similar experiment, again using the chromatography technique, and again using a fish fossil that was assumed to be 360-380 million years old. The fish was named **Dinichthys** (di-NICK-thiss) meaning "terrible fish." The researchers not only found single amino acids, but they also found

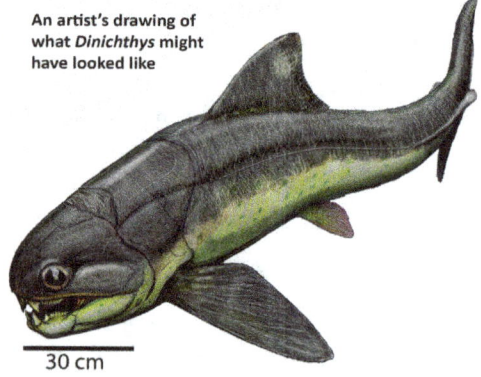

An artist's drawing of what *Dinichthys* might have looked like

30 cm

bits of **collagen**—a long, stringy molecule made of amino acids. Collagen is a very tough protein, usually white in color. Your ligaments (connecting bone to bone) and tendons (connecting bone to muscles) are made of collagen. Collagen is also found in the flexible parts of your nose and ears, as well as in many other places in your body. In fact, collagen is found almost everywhere in your body. It's what keeps you in one piece!

WHAT IS COLLAGEN?

Collagen is a protein made of individual amino acids that are fastened together to make long chains. You saw the molecular structure of amino acids on page 3. It would be very difficult to draw long strings of amino acids if you had to show every atom! Therefore, scientists represent amino acids as colored circles, sometimes labeled with single letters (G for glycine, etc.).

A string of amino acids is called a **peptide**. Special cells called **fibroblasts** make collagen peptides. The peptide strings are then braided together to make a strong "rope." This illustration shows the tight braid loosened a bit so that you can see the individual strings. **Every third amino in these strings is glycine**, the smallest amino. Because glycine is so small, this allows the braid to be very tight. (The colors in this illustration are

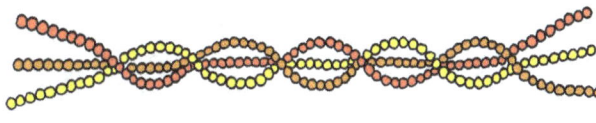

there just to make it easier to see the braided structure.)

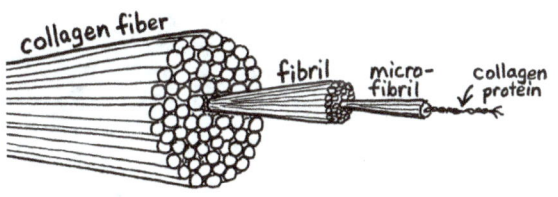

The tight collagen protein braids are then bundled together to make very small microfibrils. Then the microfibrils are bundled together to make larger fibrils. Finally, many fibrils are bundled together to make the final (very strong) collagen fibers. (Isn't it amazing that your cells know how to do this?!)

It is these collagen fibers that we can see under a regular microscope. To see the fibrils of collagen proteins, we have to use an electron microscope that uses an electron beam instead of light, and can magnify up to a million times.

Other researchers around the world did similar experiments during the 1960s and 1970s, verifying that collagen fibrils, microfibrils, or bits of braided collagen proteins, were indeed present in fossils of any age, and even in dinosaur egg shells. Other biological molecules (complex sugars) were also found. The amounts of these molecules was relatively small, but still it was surprising to find any at all since fossils were assumed to be millions of years old. However, these amazing discoveries went largely unnoticed at that time.

In 1987, another remarkable discovery was made. An employee of the Shell Oil Company had collected some bones along the Colville River in northern Alaska while he and his team were out looking for places to drill for oil. Eventually, the bones were given to a Texas museum so they could be analyzed by experts. The bones were identified as being from a hadrosaurus (a duck-billed dinosaur). When they looked closely at the bones, what they saw astonished them. These bones were not even fossilized! The preservation was remarkable—unlike any fossilized bones they had ever seen. How could they possibly be millions of years old?

duckbilled dinosaur

CHAPTER 2: Research that rocked the scientific world

Scientists continued to make surprising finds during the 1990s. In 1992, something new turned up—discovery of ancient DNA. A group of researchers were examining a termite that had become stuck in tree resin a very long time ago. At some point after that, the resin had become hard as stone. This was a great opportunity to run tests on a tiny animal that hadn't been fossilized like other animals. Being stuck in amber, the animal's cells had not been replaced by minerals.

The researchers were able to extract tiny bits of DNA and compare them to the DNA of living termites. The ancient DNA matched the modern DNA. The amber was assumed to have formed 25 million years ago, so the researchers concluded that DNA could be preserved for millions of years.

Knowing that modern termites have specialized bacteria living in their gut that help them to digest the wood that they eat, a second team of researchers extracted material from the ancient termite's gut region. (Can you imagine how tiny a termite's intestines must be?) They found short DNA sequences that matched those found in bacteria living in modern termites.

termite stuck in amber

WHAT IS DNA?

It's the very long molecule found in the nucleus (center) of every cell. It stores the information the cell needs to carry out its functions.

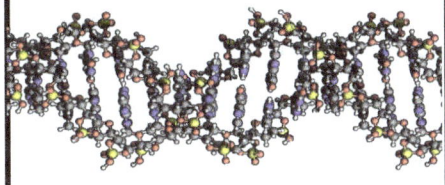

In 1995, a discovery was made that no one could have predicted. Having heard that bacterial DNA was found in the gut of a termite in amber, another research team decided to do an experiment that they probably thought would surely fail. They extracted material from the gut of what they thought was a 25 million year old termite in amber and put the ancient bacteria into a solution that was ideal for bacterial growth. They expected nothing to happen, but instead, they found LIVING bacteria swimming around in the solution! They wondered if some bacteria from the lab had gotten mixed into the solution by accident, so they ran some more tests. Their final conclusion was that the ancient bacteria had come back to life.

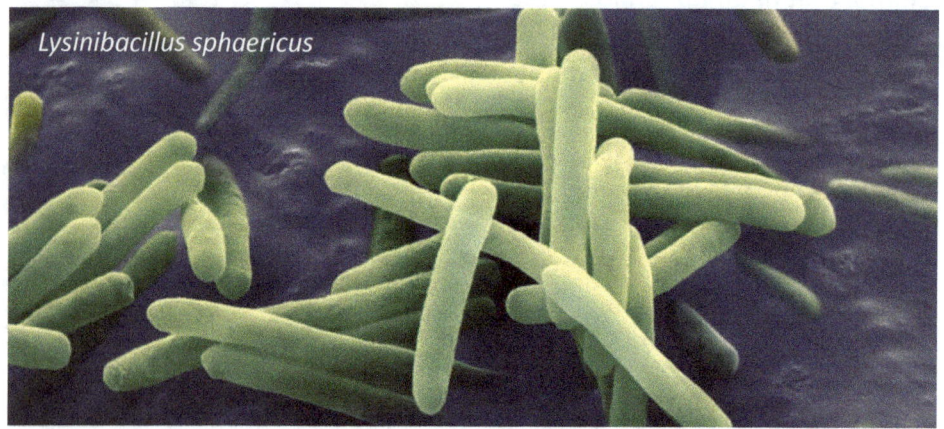
Lysinibacillus sphaericus

Some microorganisms have incredible abilities to survive harsh conditions. They can go into a form of hibernation called dormancy (almost dead but not quite) for a very long time. However, no one thought that dormancy could last for millions of years. Even thousands of years would be pushing the limit.

Now we are going to meet the first researchers that you need to remember by name: Mary Schweitzer and Jack Horner. Jack Horner was the real person behind the character of Alan Grant in the *Jurassic Park* movies. Jack was a professor of paleontology at Montana State University and Mary Schweitzer became one of his students in 1989. Jack's specialty was teaching about dinosaur fossils and he often took his students on field trips where they could learn how to dig for fossils.

One of the field trips was to a dig site in northeastern Montana called Hell Creek. Mary found out where this site got its name. Whenever they cracked open the rocks, a terrible smell came out—a smell like rotting animals. Mary thought this was a significant clue, but the paleontologists had gotten so used to it that they didn't bother thinking about it any more. However, the smell made Mary wonder if there really was rotting flesh in the rocks.

Hell Creek, Montana

After Mary graduated, she decided to continue working as a paleontologist. This was her second career. She already had college degrees in communication and education and had taught deaf students, but her lifelong interest in exploring nature had recently become a passion she felt she had to follow. One day in 1997, one of her fellow researchers in the Montana lab brought her a microscope slide to look at. She told Mary that an expert microscopist had looked at this slide of dinosaur bone and said that he was sure he had seen a blood cell. Mary took the slide to Jack so he could have a look at it. Jack took a "wait and see" attitude and told her she should try very hard to prove the bone did NOT contain blood cells. If she succeeded, then it probably didn't. However, if she was unable to prove otherwise, then perhaps this bone really did contain the remains of blood cells.

Mary and her fellow researchers did exactly what Jack had recommended and ran all sorts of tests. Chemical tests suggested that there were pieces of heme and hemoglobin in the sample. (See the next page for an explanation of these molecules.) They even took a liquid extract from the bone and injected it into rats to see if their immune systems would react to it as if it contained foreign blood cells. They did react! Of course, there weren't any truly living cells in the samples, only molecules that had been produced by the dinosaur's cells. But even to find these molecules was amazing.

WHAT ARE HEME AND HEMOGLOBIN?

Heme is a molecule found in red blood cells that can hold and carry oxygen atoms. Heme is made of atoms of carbon, nitrogen, hydrogen and oxygen, plus one atom of iron (Fe). It is the iron atom that allows heme to pick up an oxygen atom in the lungs. Heme molecules sit inside a larger molecule called hemoglobin. Hemoglobin is a protein made of long strings of amino acids that are coiled up to make just the right shape for holding 4 heme molecules.

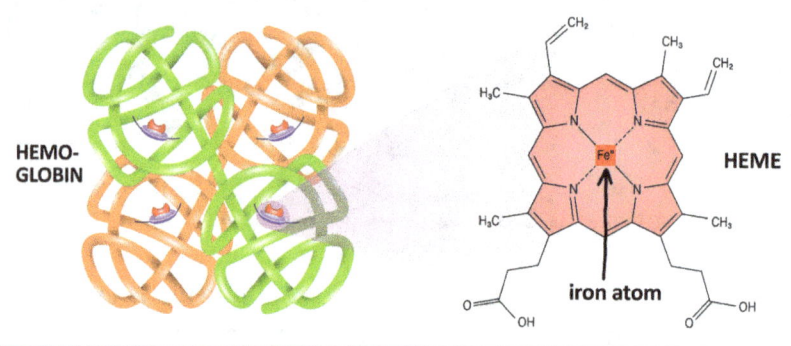

HEMO-
GLOBIN

HEME

iron atom

When Mary published the results of these blood cell experiments (which earned her a PhD degree) the reaction in the scientific community was skepticism. Some accused Mary and her team of sloppy lab procedures. After all, they knew for sure that molecules like this could never survive for 65 million years (the assumed age of the dinosaur).

Mary's next big discovery was in the year 2000. By this time she had moved to North Carolina to start a teaching job. One day, a package arrived at her lab. Someone working at the Hell Creek dig site in Montana had found a very large *T. rex* leg bone. It was too large to fit into the transport vehicle that his team was using, so they had to cut the bone in half. When they did this, the middle of the bone shattered into small pieces. The little pieces were useless for their purposes (making the bones into a museum display) but they thought that Mary might be interested in taking a look at them. When Mary opened the package and looked at them she was astonished at how unfossilized they looked. Even without a microscope she could see that some of the fragments had a special texture only seen in female birds before they lay eggs. This texture is called ***medullary bone*** and is found primarily in the femur (thigh bone). The bird's body needs a large supply of calcium very quickly in order to form egg shells. The medullary bone stores calcium for this purpose. The calcium is carried from the bone to the egg-laying organs by the bloodstream, therefore we find many

medullary bone

fossilized medullary bone

microscopic capillaries (tiny blood vessels) in the medullary bone, making it appear light red in color. When the female bird is not laying eggs, the medullary bone disappears. Male birds never form medullary bone, so the presence of this type of bone is diagnostic in determining gender. (Interestingly, egg-laying reptiles such as turtles, snakes and alligators, don't form medullary bone.)

As soon as Mary saw the medullary bone texture, she exclaimed out loud, "It's a girl!" Never before had paleontologists found a foolproof way to determined the sex of a dinosaur. Many had made guesses about body size, head shape, horns and crests. But they were guesses.

In 2004, Mary and her lab assistant made the discovery that shocked scientists all over the world. They had soaked a piece of the *T. rex* medullary bone in an acid bath in order to remove as much mineral (rock) as possible. If the fossil was nothing but solid rock, the whole thing would dissolve away. However, they had already found that the bones contained

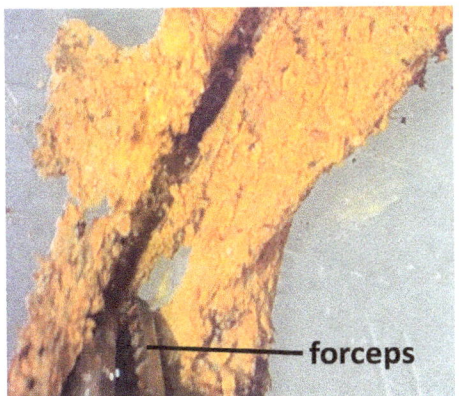

forceps

photo of similar tissue courtesy of Mark Armitage

substances that would not dissolve. They expected to find something interesting, but they were not prepared for what they saw under the microscope: something that looked and acted like fresh, stretchy collagen. Mary pulled the piece of tissue with a pair of forceps. It was soft. This was the moment that the term "dinosaur soft tissue" came into existence.

Having had so much success with medullary bone, Mary decided to try the hard outer part of the bone, called cortical bone. Her assistant put a piece of cortical bone into the acid solution for a few days giving them time to wonder what they might find.

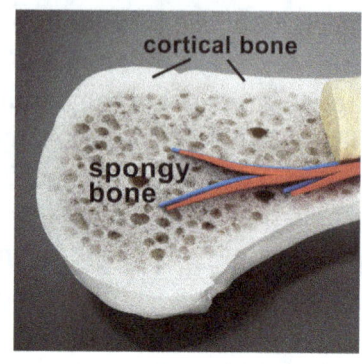

What they found was again something new. They saw what looked like blood vessels. They were in disbelief. They knew that no one would believe this unless they could repeat the experiment again and again. So they put samples from other dinosaur bones into the acid solution. Again, they found things that looked just like blood vessels. But there's more...

A few weeks later, they decalcified more bones, this time expecting to see some type of soft tissue. (The term "soft tissue" has come to mean anything other than mineralize bone. This includes amino acids, DNA, microscopic bits of protein such as collagen, all types of cells, and, of course, the large pieces of soft, stretchy tissue.) The assistant said, "Hey, look at this. I think some bugs got into our sample." But when Mary looked into the microscope, she knew exactly what those things were: bone cells called **osteocytes** (which will be explained on page 21).

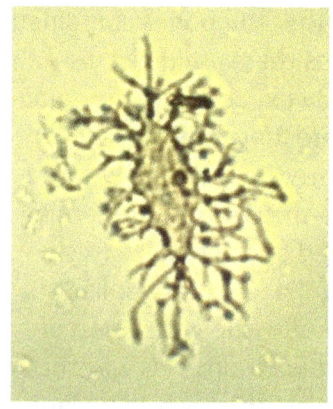

A dinosaur osteocyte (bone cell)

Mary Schweitzer and Jack Horner examine a sample of bone tissue from a T. rex (AI image)

CHAPTER 3: The controversy begins

Mary Schweitzer published her amazing soft tissue finds in 2005. This is the year that the scientific world became aware of dinosaur soft tissue. Everyone started taking sides. The finds were, indeed, quite amazing and hard to believe. Even Mary and Jack had been greatly surprised and needed to repeat the experiments many times to convince themselves that they were truly seeing what they thought they were seeing. Other scientists simply read the research papers; they never got to look through the microscope and see for themselves. Many scientists refused to believe these reports. They knew that soft tissue could not possibly last for millions of years.

Forensic scientists (who study decay rates in dead bodies) have tables and charts of the decay rates of many things, including DNA, amino acids, and collagen. They are often called in to help solve murder mysteries because they know the decay rates from hours after death, months after death, and years after death. Some of them specialize in forensic archaeology and study mummies. Mummies have been very helpful to them in their studies of the decay rates of organic molecules. Therefore, it is not just opinion about how long things take to decay; we have numbers based on observations and lab experiments. It can be proven that even under the very best preservation circumstances (very cold temperatures and very little moisture) soft tissues would have decayed away to nothing in just one

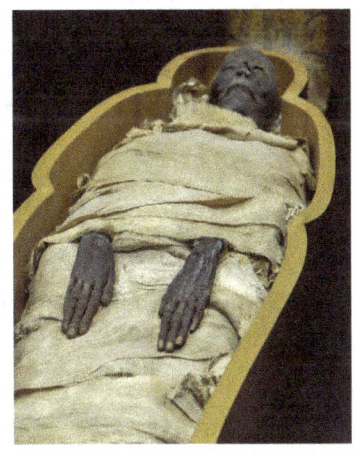

million years. Additionally, anything buried in the ground would be exposed to radioactivity. Many rocks contain uranium and thorium, unstable elements that eject harmful particles that damage and destroy organic molecules. Over millions of years, dinosaur bones would have received a considerable dose of radiation. This would not have affected the minerals very much, but would have destroyed all organic molecules.

Scientists who knew about forensic data declared that Mary must have made mistakes. Maybe the bone was not a dinosaur bone. Maybe there was contamination and something modern got dumped into the acid dish. Maybe her imagination had gotten the best of her.

Mary had anticipated the negative responses. She had taken not only photographs but also videos of her experiment. The video showed the tissue stretching as she pulled it with her forceps. You can see this video, as it is easily found on YouTube if you search for "Mary Schweitzer dinosaur soft tissue." You will have many options. Some videos are only a few minutes long rather than a full documentary.

Mary on "60 Minutes" (AI image)

You might want to watch a longer video after you read this book.

After this flash of news went through the scientific community, the media became aware of it. Unlike the response of most scientists, the media jumped on the story with glee. They were very excited to have something unbelievable to report! Mary was suddenly in the public eye. Everyone wanted to interview her. Even though she was a shy person and did not want to become famous, she agreed to be interviewed by the very popular "60 Minutes" CBS news broadcast program. Lesley Stahl interviewed Mary and they included the photographs and video of Mary's now-famous soft tissue discoveries. People who watched the news program were excited to think that perhaps scientists would soon be able to extract enough dinosaur DNA to be able to bring dinosaurs back to life. The third *Jurassic Park®* movie had been released just a few years before Mary's discoveries, so dinosaurs were very much on everyone's mind.

Not all scientists were critical of Mary. Anyone who knew her personally or had worked with her in a lab knew that she was meticulous (very careful) and never rushed to conclusions. They supported her and defended her reputation. Some of them decided to try these experiments themselves and they got similar results. The number of published papers about soft tissue began to grow. The number of skeptics shrank.

The most enthusiastic scientific supporters were biologists, geologists, and paleontologists who worked for creation-based research groups. These scientists had already come to the conclusion that the earth was not nearly as old as we are told. To them, finding soft tissue in dinosaur bones was exciting but not all that surprising.

How could anyone believe that the earth was not millions of years old? Isn't that a fact? This depends on how you interpret what we see in rocks, fossils, landforms and even in the solar system. Creationists believe that the story of the great flood that occurred during Noah's lifetime isn't just a myth. Flood stories are found in over 200 cultures around the world. Apparently, a real event like this happened within the memory of humankind. Almost every culture has handed down a narrative telling of a terrible catastrophe that came upon the earth, drowning both people and animals. If something like this really did happen, surely generation after generation would keep this story alive. As people migrated across the earth thousands of years ago, they took this story (or at least some parts of it) with them.

Several versions of the flood story include an important clue as to how it happened: water came up from underground. It didn't just leak up, it exploded with unimaginable force. Why is this detail important? A catastrophe like this would make us look at geology very differently. (Astronomers speculate that global flooding once occurred on Mars, which today has no water. Why not on Earth, which is covered with water?)

Up until the late 1700s, almost everyone assumed that the flood story was based on a real event. However, some intellectuals in the past few centuries did not like the Bible. They especially did not like the idea that humans are sinful and need to depend on God for salvation. They preferred the principles of **humanism** that originally came out of the Renaissance but were reinforced by philosophers in the 1700s. Humanism says, "Man is the measure of all things." If mankind is the measure of all things, that means that God isn't relevant. Therefore, they were very motivated to get rid of God. One easy target was to convince people that the Bible (and other ancient texts) were wrong about the history of the world. One of the easiest targets was the global flood.

A man named James Hutton, who lived in the 1700s, was one of the first writers to suggest that the earth was very old, probably millions of years old, and had never experienced a worldwide flood. He suggested that the layers we see in rocks were laid down very slowly, over millions of years, as dust and dirt accumulated at the same rate we see them accumulating today. This photo shows Siccar *(sick-ar)* Point

Siccar Point on the eastern coast of Scotland
Dave Souza, CC BY-SA 4.0, https://commons.wikimedia.org/w/index.php?curid=3831554

on the east coast of Scotland, close to where Hutton lived. Hutton could see that these rocks had an interesting history. Notice that the rocks on the right are vertical and the rocks on the left are horizontal. What happened? Hutton guessed that whatever had caused this was a process that took a long time.

After Hutton came Charles Lyell, in the 1800s, who wrote a book called **Principles of Geology**, in which he repeated Hutton's ideas as well as adding to them. Lyell's main goal was to "free geology from Moses" (Moses being the assumed writer of the book of Genesis where the flood story is found in the Bible). Charles Darwin took Lyell's book with him when he went on a long sailing trip around the world. Darwin's view of what he saw in the natural world was highly influenced by Lyell's book. The idea that rock and fossils were

millions of years old opened the door to Darwin's revolutionary ideas about evolution.

It is important to stop and note that Hutton's ideas were mostly based on his own thoughts about rocks, not on experiments. He wanted the rocks to be old, so he interpreted what he saw in terms of age. A violent, earth-shattering global flood might also be able to account for the rock layers, but this idea didn't suit him. He saw the world as he wanted to see it. We all do this, even if we don't realize that we are doing it. Scientists are not as unbiased as you think they are. Granted, they might be slightly more logical than the average person, but they are human. Their beliefs and life experiences shape their thoughts about what they see in the natural world, or under their microscopes.

The idea that rocks (and therefore fossils) were very old surged in popularity right after Darwin's book on evolution was published. The theory of evolution seemed to completely remove the need for a Creator by proposing that all living things had evolved from simple chemicals through a series of biological accidents—mistakes in DNA. Evolution says that with enough time, anything could happen—bacteria could eventually evolve into baboons. And for evolution to be possible you need millions, even billions of years. Evolution is linked to the idea of a very old earth.

(NOTE: You might already know a bit about radiometric dating. Hutton, Lyell and Darwin didn't know anything about this dating method. We'll discuss this in a later chapter.)

You can probably see where this is going. The discovery of soft tissue in dinosaur bones heated up the controversy between "old earth" (OE) scientists and "young earth" (YE) scientists. OE and YE don't always line up with belief or disbelief in a Creator. In fact, Mary Schweitzer found herself caught right in the middle of this debate. She was not an atheist. She firmly believed in a Creator, and specifically the Creator written about in the Bible. However, she also believed what she had been taught about the earth in her science classes, that the earth was very old. She found herself the object of criticism by both OE and YE scientists. The OE scientists didn't want to believe that dinosaur bones contained unfossilized soft tissues that might not be millions of years old. The YE scientists were upset that she didn't believe the Bible's account of a supernatural creation in six days and a real global flood. Mary found it hard to be scorned by both sides. She never wanted to be a controversial figure. As a result, her research went in a new direction—trying to find a way to explain how soft tissues could be millions of years old.

This chapter isn't going to resolve the controversy. We need to learn about additional soft tissue discoveries that will shed more light on the topic.

One of the best places in the U.S. to see rock layers is the Grand Canyon. Why are they so flat? Some of these flat layers cover most of the U.S. and stretch up into Canada!

CHAPTER 4: More discoveries, more controversy

Once the news about Mary's research had became general knowl-edge among scientists, lots of microscopists began to work in this field. Both OE and YE scientists went to work. In the decade following Mary's 2005 paper, there were at least 50 published papers about all kinds of preserved soft tissues. Remember, the term "soft tissue," includes a wide range of tissue types, anything from simple amino acids and tiny proteins to actual soft tissue like Mary found. This term has caused some confu-sion because it includes things that don't seem to us like tissues. For ex-ample, a scientist in 2012 found molecules of black pigment in a fossilized cuttlefish right where its ink sac used to be. The mol-ecules were not soft and squishy, of course, but they were classified as a soft tissue find. In 2013, a scientist found a fossilized mosquito that contained pieces of heme, undoubtedly the remains of its last blood

a cuttlefish

meal before being fossilized. As discussed on page 10, pieces of heme or hemoglobin can also be counted as soft tissue discoveries. Recently, there has been a transition from the term "soft tissue" to the more accurate "original biomaterial." We'll switch over to this term.

Some scientists found tissues more similar to what Mary found. In 2006, a scientist found preserved bone marrow in a partially fossilized frog from Spain, assumed to be 10 million years old. The marrow inside the bone was still red and yellow, just like the marrow found in living frogs. There were also more discoveries of truly soft tissues like stretchy collagen, but in the world of academia there's a rule that only the first report gets published by leading journals. If you want to get published, you have to find a new spin on your discovery, something different about your find, before a journal will consider it. We don't know how many other scientists found truly soft tissues but didn't get their discoveries pub-lished. However, we do know that there were enough of these reports that people stopped questioning Mary's finds and just accepted them.

However, the finding of original biomaterial in fossils hadn't completely ceased to be controversial. The next person to get caught in the middle of the controversy was Mark Armitage, an expert microscopist who had been hired by a university in California to install and run some very expensive electron microscopes in their lab. (We already noted that although many biomaterial samples can be observed in regular microscopes, there are situations in which electron microscopes can be very useful.) Mark was good at his job. All his coworkers respected him and appreciated both his technical skill and his patience with those he trained to use the equipment. The professors in his department encouraged him to go ahead and use the equipment to do his own research projects.

Mark had done a variety of research projects in the past, but had recently become interested in the new field of original biomaterial in fossils. In 2012, he decided to dig some bones himself (with some help) so that he would know exactly what conditions they were found in, and could make sure the samples were properly handled and preserved. So where to go? Hell Creek, of course. The stinkier the better!

At Hell Creek, you hardly have to dig. Sometimes bones are just sticking out of the ground. At worst, you only have dig a little bit, and the dirt around the bones is relatively soft. In some places in the world, you have to spend weeks chipping bones out of rocks. But not here. Mark soon found what would become one of the most famous Hell Creek fossils: a triceratops horn that was only partially fossilized. In fact, if it had been

a cow horn, it would not have attracted much attention at all. It would have been just a dirty old bone. However, this area was a dinosaur grave-yard, and this horn was quite obviously a triceratops horn.

Mark and his crew were quite excited about this find, but they had to be careful getting it back to the lab and preparing the samples. Even though it had been over 4 years since Mary Schweitzer's shocking soft tissue discovery, some scientists were still very skep-tical, and if mistakes were made in the lab, the research would be dismissed as faulty.

Mark

WHAT IS AN OSTEOCYTE?

An osteocyte is a poor little cell that got left behind. Really, it is! Bone tissue is made by cells called **osteoblasts**. They move along together, producing a network of collagen fibers that they then stuff full of minerals such as calcium, magnesium and phosphorus. They get these minerals from the food you eat. The osteoblasts (show in orange) are constantly laying down new bone, but as they work, some of them stay behind and get stuck in place, turning into osteocytes (shown in yellow). While they are turning into osteocytes, they grow lots of skinny "arms " called **filopodia**. Their filo-

podia reach out and touch the filopodia of neighboring osteo-cytes. The filopodia are like hollow tubes that water, nutrients and chemical messag-es can flow through. Osteocytes monitor the bone's health and can signal other cells, telling them how to keep the bone from becoming weak.

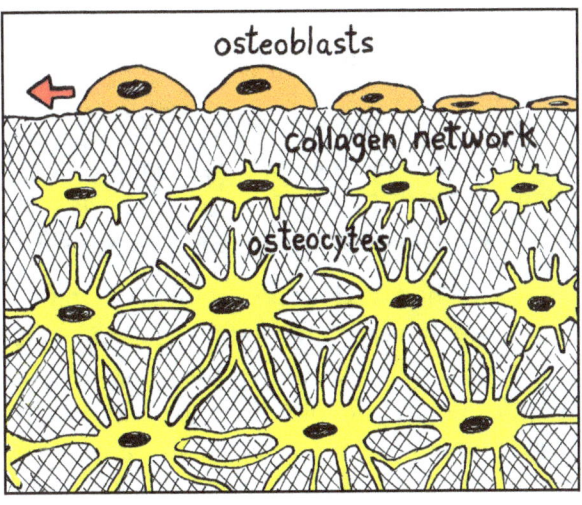

IF WE ZOOM OUT, WHAT DO OSTEOCYTES LOOK LIKE?

It's interesting (and important) to know what osteocytes look like if we zoom out a bit and see the bigger picture of bone tissue. The osteoblasts (the ones who create the collagen network and stuff it full of minerals) go around in circles. They start with a small circle that goes round a grouping of an artery, a vein, a nerve, and a lymph vessel. As they go around and around, the circle grows larger. Then they stop making that circle and start making a new one. They do this constantly. Right now, your osteoblasts are creating these circles. If you are young, they are working faster than if you are older, and if you break or fracture a bone they have to speed up in order heal it quickly. Newer circles are built right on top of older ones. In this diagram, you can see the really old ones are barely visible.

If another part of your body needs more calcium (calcium is needed for many bodily processes, such as muscle contraction) there are cells called **osteoclasts** that can go back and dissolve the network that the osteoblasts made, and get the minerals back out again and put them into the bloodstream.

At this level of magnification, the osteocytes look tiny. Their filopodia made them look like ants. The red things are blood vessels.

Mark had read reports of original biomaterial in dinosaur bones, and it was exciting to think that he now had an excellent tissue specimen to examine. It was obvious that the bone tissue inside the triceratops horn was not completely fossilized. He used several microscope techniques to examine the bone. He took both flat, transparent images (using the kind of microscope you are familiar with) as well as 3D images with electron microscopes. When he saw the "bugs" he knew exactly what they were.

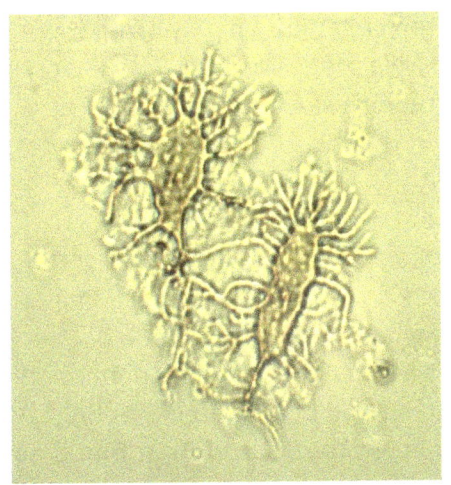

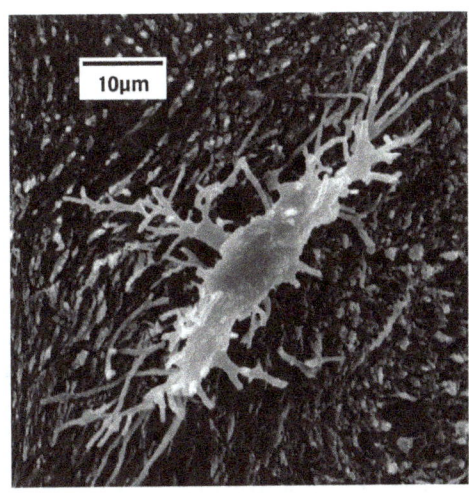

These two osteocytes were connected through their filopodia when they were alive. The bone matrix around them was dissolved either by time or by the acids that Mark used to treat the samples.

This image shows you another view of an osteocyte. The black background is remaining bone that did not dissolve. The white specs on it are bits of collagen. The white rectangle with "10μm" is a measurement key.

Osteocytes are very small. The measurement key in the black and white image above reads "10 micrometers" (or "microns"). That u-shaped thing is the Greek letter "mu" (their "m"). Bacteria are 1-3 micrometers. Your red blood cells are 7-8 micrometers. Human hairs are about 80-100 μm wide.

Mark and his co-author submitted the findings to a well-respected research journal called *Acta Histochemica*, and they accepted his article for publication in their next issue in February of 2012. On the day that this issue became available online, one of Mark's coworkers burst into the lab and began angrily shouting at him. "We're not going to tolerate your religion or your creationist projects in this department! This is a science lab!" Mark hadn't mentioned religion or creation in that paper. The paper simply reported on what he had found in the triceratops horn. People had been publishing papers on biomaterial since 1954.

The United States has laws against discrimination because of personal beliefs. When Cal State hired Mark, they knew he was a Christian. That was fine with them. After all, Christianity teaches that we should be honest, patient, kind, and hard-working. All of these qualities make good employees.

Mark knew that his rights had been violated so he reported this incident to his supervisor. "Not to worry," his supervisor said. "Your work is excellent. Very sorry for this person's rudeness. Just ignore him." The supervisor failed to warn this rude person that his behavior had been not only inappropriate but also illegal.

Unfortunately, soon after this event, Mark's supervisor left his job and his replacement was... the very man who had angrily burst into Mark's office. One of this new supervisor's first actions was to fire Mark. He didn't say why. Mark did a little research and found out that a departmental committee had met in secret the very day that the paper had been published online, and this rude person had been in that meeting demanding that Mark be fired. He was not yet the supervisor at that time, so he had no authority to hire or fire anyone. But as soon as he rose to the level of supervisor, he was able to take action.

Because the firing had been illegal, a law firm was glad to take up the case and they easily won. The judge ruled in Mark's favor. However, Mark did not get his job back. He continued to research biomaterial in fossils using a different lab located in Arizona. Then, several years later, he was able to strike out on his own and set up the organization that he currently works under, the Dinosaur Soft Tissue Research Institute, in Washington state.

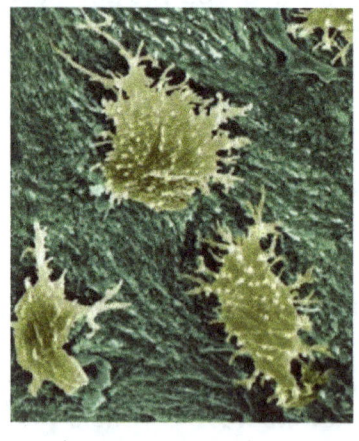

Another image by Mark Armitage. Sometimes color is added to make the image more appealing to viewers.

Neither Mary nor Mark had ever imagined they'd be the center of controversy. Both are shy scientists who want nothing more than to spend hours in front of their microscopes, then share their exciting finds by publishing papers with descriptions and photographs. But their stories don't end with these chapters. They will both continue to be important players in the unfolding story of dinosaur soft tissue.

CHAPTER 5: The research race

After 2005, a race began. As previously noted, journals will only publish a research paper if it announces a new discovery. Scientists are always welcome to repeat experiments in order to verify them, but they won't be able to get them published in a journal. Why the need to publish? Often, their jobs depend on it. Researchers must constantly get grants (free money handed out by universities or large corporations) in order to stay employed. Original biomaterial in fossils was a still a relatively new field with lots of topics still available, so it was a goldmine of grant money. There were many fossilized animals and plants to choose from. Besides the many types of dinosaurs, there are also well-preserved fossils of ancient turtles, amphibians, birds, fish, plants, and ocean invertebrates (things like clams, sea sponges, and squid). Everywhere they looked, scientists were finding biomaterials.

Mary Schweitzer stayed in the race and published papers about chains of amino acids in *T. rex* bones, collagen and blood vessel proteins in hadrosaurs, proteins in sauropod eggshells, even more types of proteins in *T. rex* bones, pieces of DNA in *T. rex* bones, pigment molecules in an extinct bird, proteins from the claw of an oviraptor, muscle proteins from sea turtles, flexible skin and fat cells from an **ichthyosaur** (similar to our toothed whales today), and what appeared to be chromosomes (bundled up DNA) inside cartilage cells of a baby duck-billed dinosaur. Mary wasn't always the lead researcher on these projects, however. After graduation she became a professor, like Jack Horner, and had graduate students working under her, doing a lot of the lab procedures.

Ichthyosaur fossil

Mark also stayed in the race. (There were many other scientists doing biomaterial research, too, but mentioning all their names would be too much information. If you want to see more names, check out the reference in the bibliography.) He went back to Hell Creek, of course, but also to some sites in Texas and Oklahoma. He acquired specimens of nanotyrannus, Edmontosaurus (a type of hadrosaur), *T. rex*, a tiny amphibian called *Cacops* (which means "looks ugly"), and dimetrodon.

Cacops

digging for Edmontosaurus bones

One of the first things Mark set out to do was to help put an end to the claims that these soft tissue cells were nothing more than "biofilms." Bacterial cells can cooperate with each other in order to survive. They can get together and secrete a slimy film made of sugars, protein and fats. This film can stick very tightly to a surface. Biofilms can be quite tough. The white plaque at the base of your teeth is an example of a biofilm. The dentist must use a special tool to scrape it off because a toothbrush isn't strong enough to remove it. Perhaps the osteocytes were not really remnants of the actual cells, but just a biofilm that looked like cells. Mary Schweitzer had already done some experiments to prove that these were not biofilms but Mark added to the argument. The tissue in the triceratops horn was layered, just like in living horn tissue. Mark pulled layers one by one and saw osteocytes in every layer. You will remember (from page 21) that osteocytes have delicate "arms" called filopodia that reach out and touch the filopodia of all their neighbors from side to side and also up and down. Mark was looking at a 3-dimensional network of osteocytes. It is impossible for biofilms to create 3-dimensionally stacked layers. There has been enough evidence against the biofilm theory that no one really advocates for it anymore.

Next, he set out to disprove one of Mary's theories about how these soft tissue structures could survive for millions of years. Her proposal was that the iron inside the dinosaur's red blood cells was released soon after the dinosaur died and the iron from the heme was able to cause chemical reactions in the tissues that preserved them. As you saw in the picture on page 22, there are many tiny blood vessels present in bones. They are located in the middle of those circular layers that osteoblasts make.

Mary designed an experiment using blood from living chickens and ostriches. She had to be very careful to prevent the blood samples from clotting. (You have seen blood clotting. Any time you cut yourself, blood leaks out and immediately a complicated chemical reaction occurs so that the blood cells stick together to stop the bleeding by forming a soft clot. As time goes by, the clot dries out and hardens into a scab.) The experimental process involved using a chemical to prevent the clotting process, then using a centrifuge to spin the blood samples at very high speed in order to separate the red blood cells from all the other cells and fluids in the blood. Then they had to cut open the red cells to release the iron-containing hemoglobin. (Remember, it was the iron that they really wanted.)

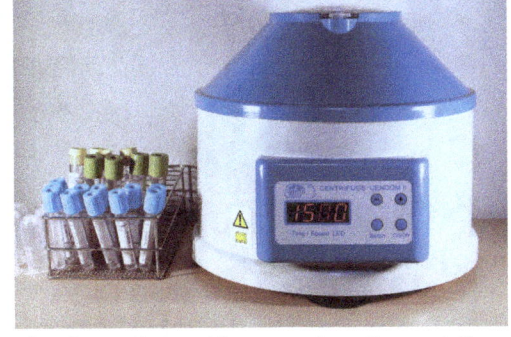

Blood samples waiting to go into the centrifuge

When recently dead bird bones and blood vessels were soaked in a heme solution for up to two years, she did see improved preservation of the birds' collagen and other biomaterials, compared to dead birds that were not treated. However, does this explain preservation over millions of years? Mark began to investigate clotting in dinosaur bones.

A little research into iron reactions revealed that iron can also cause a lot of damage. Iron and oxygen are very corrosive, as you know from watching rust form on metal. Iron released from heme might just as well destroy tissues as preserve them. Then there was the question of whether the blood could leak out into the bone tissue before it clotted and became stuck inside the vessels. As you can see in the illustration on page 22, the bone surrounding the central blood vessel is very compact and hard. Would the iron be able to reach the osteocytes?

Mark still had some triceratops horn tissue left, plus he had acquired some ribs and vertebrae. To prepare the bones for viewing, he had to "thin section" them, cutting them into very thin slices. Light has to be able to penetrate up though your sample. Everywhere he looked, Mark found what looked like blood clots inside vessels. The images shown here are actual pictures that Mark took. But were they really clots? One way to find out was to use a technique involving ultra-violet light. Some substances will glow when exposed to UV light. Iron happens to glow with a light blue color. You can see in the blue image below that the vessel is full of iron (light blue), but no iron is present in the surrounding bone tissue. Heme and iron did not leak out of the vessels. So much for the iron theory.

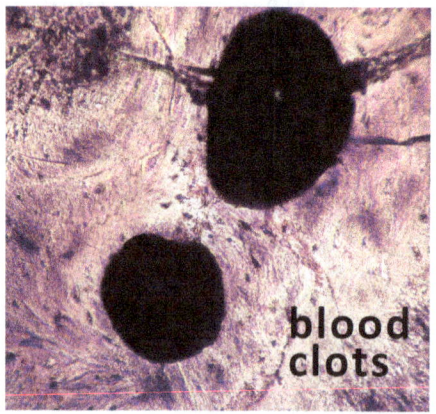

blood clots

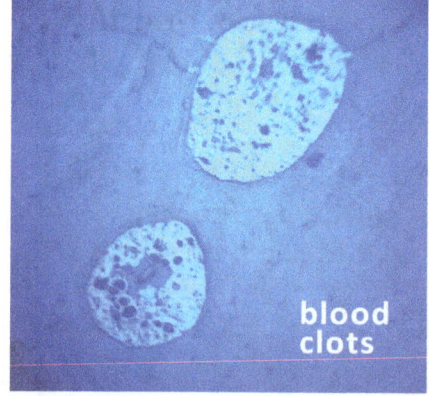

blood clots

Mark found evidence of clotting in other bones, too— hadrosaurs, dimetrodon, *Cacops*— the assumed age of these fossils ranged from 65 million to 250 million years old. Was there a significance to all this clotting? He did some research and found out that severe clotting occurs in the following medical situations (seen in humans): severe injury, infections of the blood, some types of cancer, complications in childbirth, Covid-19 infections, and drowning. Which is most likely for these ancient animals? The clots were everywhere, in bones from different body parts, so injury is unlikely. Blood infections are uncommon. The option that makes the most sense is a massive drowning event. The Hell Creek area is dry now, but the landscapes was quite obviously carved out by water in the past. That might explain all the bones from Hell Creek, but what about the bones from Texas?

Clots were only one of the many discoveries Mark found in the bones. He also found what appeared to be blood vessels, and even tiny valves that had once been inside a vein. The most spectacular find was the nerve fibers because they shimmer with color if you view them using polarized light. There was absolutely no doubt that these were nerve fibers because there isn't any other type of tissue that has this cross-hatching pattern. The pattern is due to special tissues that wrap around the fibers.

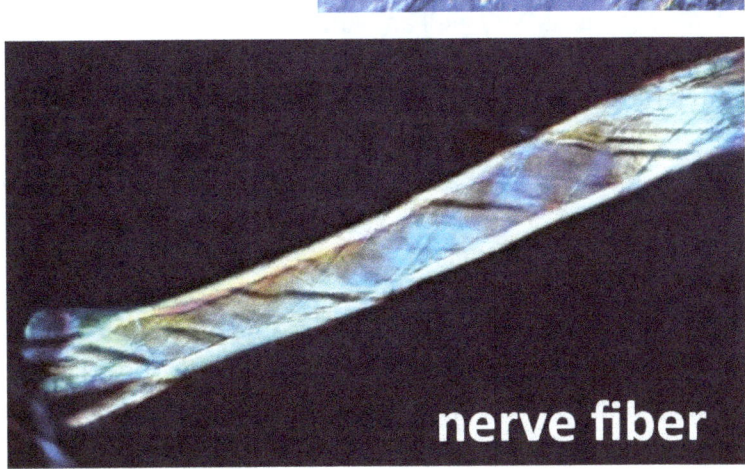

vessels

nerve fiber

Then Mark decided to look for nerves in other fossils. He acquired some *Cacops* bones from a site in Oklahoma. The assumed age of *Cacops* is about 270 million years, which is more than 200 million years older than Triceratops. Would he still find nerves in a specimen that is assumed to be this old?

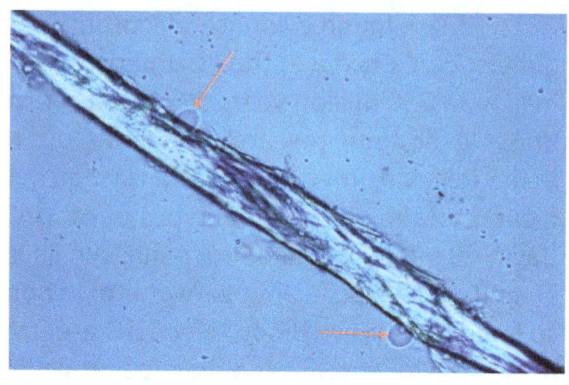

Lipid droplets that have oozed out of Cacops nerves

Not only did he find nerves, he found something even more surprising. He saw what looked like tiny droplets of oil (lipids) oozing out of some of the nerve fibers (pointed out by the red arrows). Nerve cells are covered with a fatty substance called **myelin** which helps to insulate the nerve from its surroundings. The myelin functions a bit like the rubber coating on an electrical cord. The insulation prevents electrical signals from leaking out of this biological wire.

How did Mark know these were lipid droplets? There are only a limited number of possibilities. They couldn't possibly be water because water would have evaporated not long after the fossil began forming. They weren't anything that Mark had put onto the microscope slide. He knew all his chemicals very well and knew what each one would do or not do. He had used these chemicals with thousands of dinosaur tissue slides. Also, when he pushed down on the thin piece of glass covering the nerve (the coverslip), he could see the droplets coming out of the nerve.

Many other scientists beside Mark Armitage and Mary Schweitzer are actively researching original biomaterial in fossils; it's just not possible to mention all of them in this short book. Every day there are new reports being published. Twenty years ago, some paleontologists were still denying that these reports could be true, but now everyone accepts them. They are no longer controversial. What is still controversial is how to interpret what we are seeing.

CHAPTER 6: Why millions of years?

Discoveries of original biomaterial in fossils continue to be made, but they are getting less dramatic (so far). We've already seen amino acids, tiny proteins, pigments, stretchy collagen, osteocytes, nerve cells, blood clots, and blood vessels. Additionally there have been reports of actual blood cells as well as positive test results for the presence of DNA. New reports tend to be about the same structures turning up in more kinds of fossils. It doesn't seem to matter what the assumed age of the fossil is. Mark found delicate nerve fibers in bones that ranged in assumed age from 65 to 270 million years old. Shouldn't the older bones have fewer structures? Didn't time make any difference at all? Why do many people think the dinosaurs are millions of years old, anyway?

We've already learned that the idea about rock layers forming over millions of years started in the 1700s, before the theory of evolution ever came on the scene. People like Hutton and Lyell could see fossils buried in these layers of rock, so their natural conclusion was that these animals must have been the same age as the rock around them. It seemed that biology and geology were linked somehow.

It was Darwin, in the 1800s, who put forth a theory about why particular animals occurred in particular layers. He proposed that over millions of years, simple animals had evolved into larger, more complex animals. He suggested that the climate of the earth had changed dramatically many times, causing animals with better adaptations to survive the changes. But in the 1800s, scientists had not yet assigned ages to all the layers.

In the early 1900s, great strides were made in physics and chemistry. Radioactive elements had been discovered even before the turn of the century, and now they were discovering many types of unstable atoms. To understand unstable atoms, we need to review the basic structure of atoms.

Inside the nucleus of each atom are protons and neutrons. Both of these are made of even smaller particles called quarks. Quarks have strange names. (In fact, "strange" is the name of one type!) The quarks in protons and neutrons are called **up quarks** and **down quarks**. A proton has two up quarks and one down quark. A neutron has two down quarks and one up quark. It appears that this ratio can change: a proton can turn into a neutron, or a neutron can turn into a proton. This is one of the keys to understanding unstable atoms.

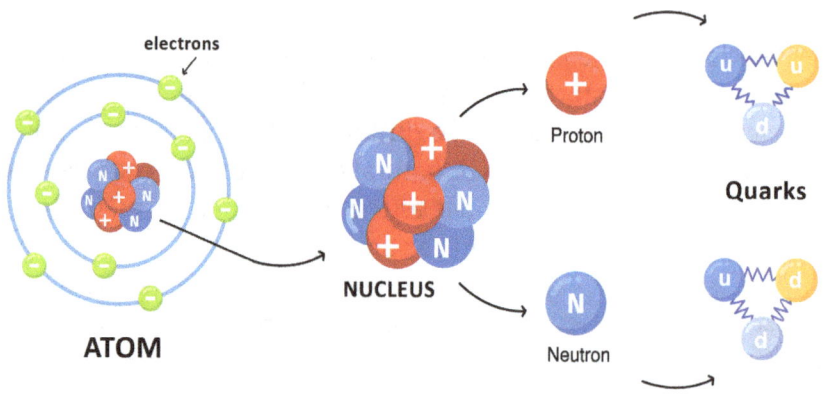

The identifying mark of an element is its number of protons. All carbon atoms have 6 protons. All nitrogen atoms have 7. Helium atoms have 2, and poor little hydrogen has only one. You can tell how many protons each type of atom has by looking at a Periodic Table. The number of the element is the number of protons you will find in that atom's nucleus. If the number of neutrons changes, this does not affect the identity of the element. Most carbon atoms, for example, have 6 neutrons along with the 6 protons. However, if the number of neutrons increased to 7 or 8, the carbon atom would still be carbon. The atom would be a bit heavier, but it would still be carbon.

Another key to understanding unstable atoms is their size. Larger atoms have larger nuclei and many more neutrons. For example, uranium is number 92 on the Periodic Table so it has 92 protons in its nucleus. It usually has 146 neutrons also, giving it a total weight of 238. This is a very large atom. Large atoms tend to fall apart much more easily than smaller ones. Imagine two cookies, one small and one very large. Which one is more likely to fall apart when you transfer it from the pan onto your plate? Large atoms are also more likely to crumble. One way an atom crumbles is to expel neutrons or pro-

Uranium is a metal.

tons from its nucleus. Very often, a group of 2 protons and 2 neutrons are expelled together. But wait—if an atom does this, won't expelling 2 protons change its identity? Yes, it will.

When uranium loses two protons it changes into element number 90, thorium. But thorium is still a pretty big atom and it will eventually do the same thing, turning into a smaller atom. This continues until the size of the atom is small enough to be stable and stop spitting out parti-cles. The stable atom that uranium finally turns into is number 82, lead. This process of getting smaller is called ***radioactive decay***. (Elements that spit out particles are radioactive. Radium was one of the first radioactive elements discovered, and its name came from the "rays" of energy it was spitting out. These rays are beyond the scope of this book.)

Scientists noticed that some rocks contain both uranium and lead atoms. This is especially true of ***igneous*** rocks that were once molten (hot liquid) and cooled over time. These rocks often have tiny crystals called ***zircons***. Zircons are very hard and therefore not very much affected by the environment around them. If uranium is trapped inside, then we should be able to see many of the smaller decay products, like thorium and lead. (Many zircon crystals are microscopic, like the crystal shown here. This one is only 250 µm long.)

Because the decay pattern of uranium is known, we might be able to calculate the time since that zircon crystal formed by counting those atoms that came from the decay. Now the math on this will sound a bit strange. The calculated time that it would take for an average uranium-containing rock to have half of its uranium turn into lead can be as high as 4.5 billion years. Obviously, this has never been observed by anyone. Mathematicians can do something called *extrapolation*. They can take just a little bit of actual data, put it on a graph, then extend the lines of the graph to see where the line would go if the numbers kept getting higher. (This is an over simplification, but gets the idea across.) Thus, if we measure the amount of uranium and lead in a crystal, we can get an estimated age up to 4.5 billion years. This has actually been the estimated age of the earth since 1956, when this technique was developed.

In recent decades, other techniques have come into play. It turns out that other unstable elements can be used in this way. A very popular method has been the potassium-argon method. The symbol for potassium is K, and argon is Ar, so we'll switch to those and you can just say "potassium" when you see the letter K.

Potassium always has 19 protons, but can have 19, 20, or 21 neutrons. This makes total weights of 39, 40, and 41. We call these different types of potassium

Feldspar is a silicon-based mineral that often contains potassium.

"isotopes." Isotopes are the same atoms with different numbers of neutrons. K-39 is the most common, and most stable, of these isotopes. K-41 is less common but is also stable. These atoms seem to be happy with their number of neutrons. However, a very small fraction of the potassium atoms in the world are K-40. This number is not stable, and eventually this atom will capture a free electron, allowing one of its protons to turn into a neutron. One less proton means that the atom is no longer number 19 but number 18. Potassium is a solid, but argon, number 18, is a gas. A solid turning into a gas? Yes, the world of chemistry can be weird! If K-40 decays into Ar-40, we suddenly have a gas molecule trapped inside a solid rock.

To use the K-Ar dating method, the rock sample is crushed and the amount of Ar-40 is determined. They don't directly measure the K-40 because the amount is so small. Only about .01% of all potassium is K-40. So instead, they measure the K-39 and then calculate what the amount of K-40 should be assuming the ratios hold true inside this rock. Then these numbers are put into a mathematical formula and out pops a number that is the estimated age of the rock. This number is always in the millions or billions.

The radiometric technique can be used with other elements that have at least one unstable isotope. They can even use just argon, comparing Ar-40 to Ar-39. The Ar-Ar dating method is currently one of the most accepted and trusted dating methods.

This is how igneous rock (cooled lava) is dated; but fossils are not found in igneous rocks. The type of rocks in which you find fossils really can't be dated. In many places, however, there are stacked layers that have igneous rocks between sedimentary rocks. So they assign dates to those igneous layers and assume that everything below that layer is older and everything above that layer is younger. Additionally, you assume that the **sedimentary layers** (sandstone, limestone, shale) formed at the same rate back then as they do now, extremely slowly. A layer of limestone of a certain thickness is assumed to be a certain age because of how thick it is.

Then you find similar layers in various places around the world to collect more data. Similar fossils are taken as markers of particular layers. If you find a *T. rex* fossil, that layer of rock will be assigned to the Cretaceous *(kree-TAY-shus)* time period of 66 to 69 million years ago. *T. rex* acts as an indicator fossil. They won't even bother dating the rock around a *T. rex* fossil. They assume that the rock is 66 to 69 millions of years old.

There isn't a single place in the world where you can see all the geological layers stacked up from the bottom to the top. The charts of all the layers are a compilation of data from around the world. The igneous rocks gave the original millions of years dates. This chart does not indicate where the igneous layers occur because they occur in different places around the globe.

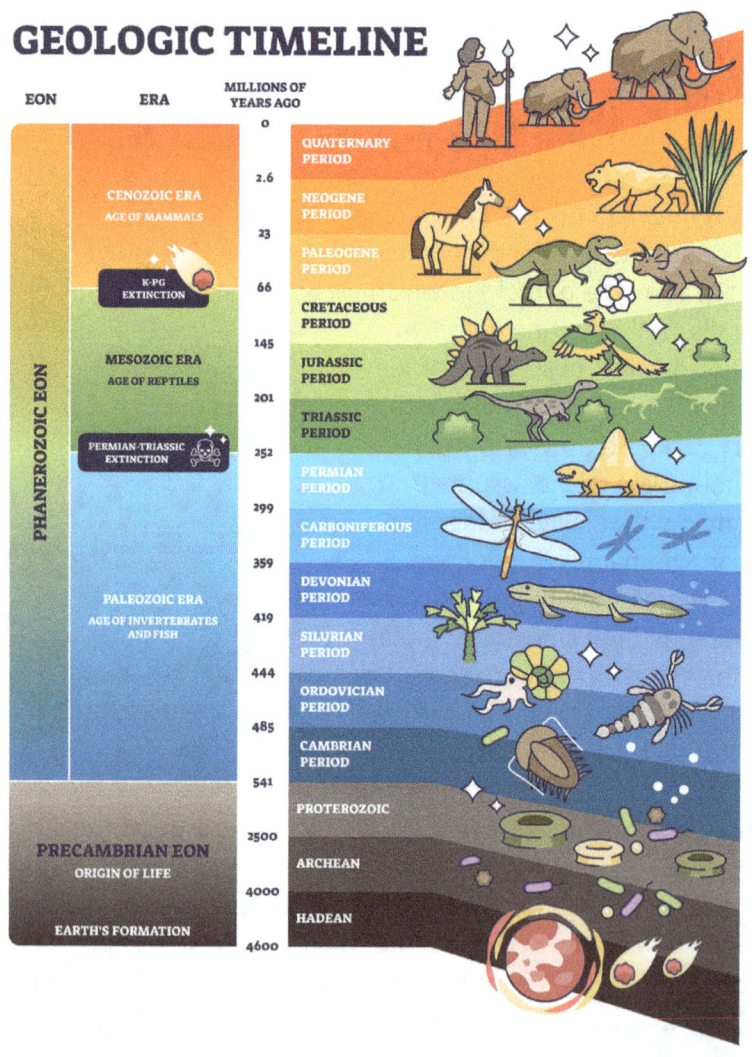

GEOLOGIC TIMELINE

EON	ERA	MILLIONS OF YEARS AGO	
		0	QUATERNARY PERIOD
	CENOZOIC ERA AGE OF MAMMALS	2.6	NEOGENE PERIOD
		23	PALEOGENE PERIOD
	K-PG EXTINCTION	66	CRETACEOUS PERIOD
PHANEROZOIC EON	MESOZOIC ERA AGE OF REPTILES	145	JURASSIC PERIOD
		201	TRIASSIC PERIOD
	PERMIAN-TRIASSIC EXTINCTION	252	PERMIAN PERIOD
		299	CARBONIFEROUS PERIOD
		359	DEVONIAN PERIOD
	PALEOZOIC ERA AGE OF INVERTEBRATES AND FISH	419	SILURIAN PERIOD
		444	ORDOVICIAN PERIOD
		485	CAMBRIAN PERIOD
		541	PROTEROZOIC
	PRECAMBRIAN EON ORIGIN OF LIFE	2500	ARCHEAN
		4000	HADEAN
	EARTH'S FORMATION	4600	

CHAPTER 7: Why thousands of years?

Radiometric dating is usually treated as the final word on the age of rocks and fossils. How can you argue with lab results? But if you take a closer look, you might be surprised to find out just how many assumptions are made before any testing is done. The following list of assumptions was taken straight out of Wikipedia, a source that strongly supports the old age theory. Here are four statements about the K-Ar dating method that must be assumed to be true:

1) K-40 decays at a steady rate (the same as it does now) and is not affected by the environment around it.

2) The K-40 to K-39 ratio has always been the same as it is now.

3) All argon gas in a sample was produced by K-40.

4) The sample has been "closed" since its formation, and no argon gas has leaked in or out. (The rock was never re-heated, which would have allowed some of the argon to escape.)

None of these statements can be tested. No one was there for all those years watching the rocks, checking to see if anything was leaking in or out. No one was there long ago to see if K-40 used to decay faster or slower. We don't know if a sample experienced re-heating.

Scientists do try to be very careful in their procedures, trying to think of ways to do testing without having to make assumptions. In fact, a very clever scientists found a way to use just argon atoms. This avoids the problem of having to use two different machines, one to measure solid potassium and one to measure gaseous argon. One machine is used to measure the levels of three isotopes of argon. First, they put the sample into a nuclear reactor and bombard it with free neutrons. Some of these will strike the nuclei of the potassium atoms and knock out a proton.

An AMS machine like this is used to count argon atoms.

With one proton missing, these potassium atoms turn into Ar-39. Now they measure the levels of three Ar isotopes: Ar-36, Ar-39, and Ar-40. They do this by slowly heating the sample while it is inside a machine that can detect and count all the argon atoms that escape. Every atom can be counted. Most of the time, the dates given by K-Ar and Ar-Ar are similar.

The Ar-Ar method does get around a few assumptions necessary for the K-Ar method, but there still is one primary assumption: that the levels of Ar-36 and Ar-40 in the atmosphere were the same in the past as they are today. The levels of argon gas in our atmosphere right now are used as the "normal" that the rock samples are compared to. What if the levels of argon in the past were NOT the same as they are today? What if there was a major catastrophe that changed the earth's atmosphere? What if an earth-shattering geological event caused the release of an un-

usual amount of free neutrons into the earth? (This would be similar to putting the rocks into a nuclear reactor.) These factors would change our interpretation of the data. The age estimates might go way up or way down.

At this point you may be thinking, "I thought this book was about dinosaurs. It seems to have turned into a physics book." We're going to get back to the topic of dinosaur bones soon, but there are still a few more important things to learn about dating methods. There is actually a way to test the dinosaur bones directly and not the rock layers above or below them. Because living things contain carbon-based molecules, we can test carbon atoms instead of uranium, lead, potassium or argon.

Let's look again at the diagrams of amino acids. Notice all the C's. Glycine has 2, and alanine has 3. Some aminos have many more. As we learned, C stands for carbon.

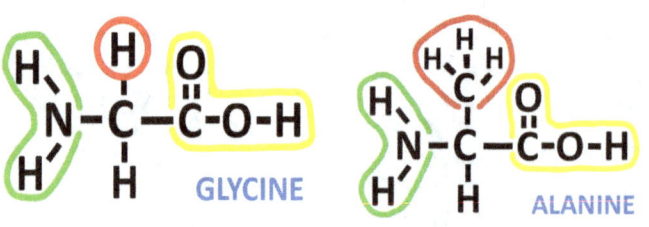

GLYCINE

ALANINE

Collagen is made of amino acids, and so are the many other types of protein that make body structures. DNA is based on carbon atoms. If we can use carbon as a dating method, we should be able to date the remaining tissue in the bones.

Once again, we are going to have to talk about *isotopes*. Carbon always has 6 protons and usually has 6 neutrons. Add 6 plus 6 and you get 12. A normal carbon atom has a weight (or, more technically, a "mass") of 12, so normal carbon is C-12. (You can say, "C twelve" or "carbon twelve." Either is fine.) A small percentage of carbon atoms have an extra neutron and are identified as C-13. Another small percentage of carbon atoms are C-14, but this form of carbon has a story behind it.

As far as we know, the only natural source of C-14 is the upper atmosphere. High-energy cosmic rays from the sun strike the atoms in the atmosphere. Sometimes they hit a nucleus and cause a neutron to go flying off. If that loose neutron happens to hit the nucleus of a nitrogen atom (and most atoms in the atmosphere are nitrogen) it can dislodge a proton and take its place. Nitrogen atoms normally have 7 protons and 7 neutrons, giving it a mass of 14. But this loose neutron changes the count to 6 protons and 8 neutrons. The mass is still 14, but now that the atom has 6 protons, its identity changes to carbon. Thus, we have an atom of C-14.

This C-14 atom behaves like a C-12 atom and can combine with two oxygen atoms to make carbon dioxide, CO_2. The carbon dioxide goes into the lower atmosphere and is taken up by plants who use it for photosynthesis. Then animals come along and eat the plants, so they ingest both C-12 and C-14.

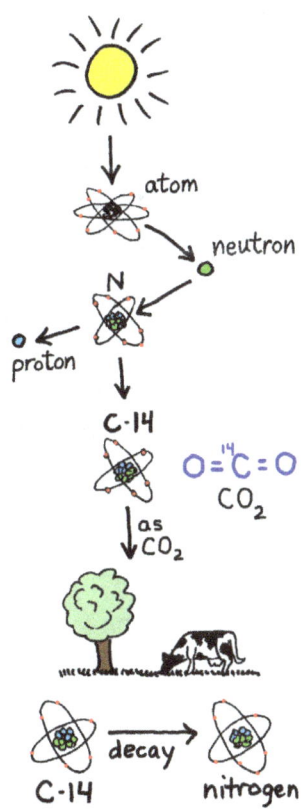

Our bodies have billions of C-14 atoms in them because we have eaten plants all of our lives. Even if you don't like fruits and vegetables, you still eat plants. That donut you ate was made from wheat, which is a plant. Everything would be fine if those C-14 atoms just stayed as C-14. But remember their history? They used to be nitrogen atoms. They really can't stay carbon atoms forever. At some point, they can and do turn back into nitrogen. Your collagen, your proteins, your DNA all have C-14 in them. Then, one day, the extra neutron in the C-14 turns back into a proton and suddenly the carbon atom is gone and there is a nitrogen atom in its place. Does this create problems? You bet! This is one of the causes of aging. Your

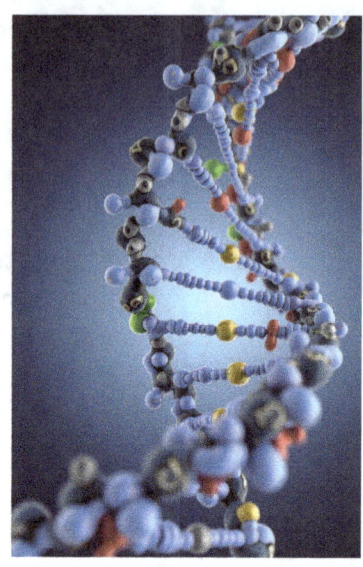

Some of these "beads" in this DNA model represent carbon atoms.

body molecules don't work so well after lots of carbons have turned into nitrogens. Your cells do have repair mechanisms, but even those mechanisms have C-14 in them!

This is how C-14 can be used to date dinosaur bones. As long as an animal is alive, it keeps taking in a fresh supply of C-14 every day as it eats. But when the animal dies, its intake of C-14 stops. Death starts the C-14 clock ticking. After about 5,700 years, half of the C-14 atoms in the animal's dead body will have turned into nitrogen. In another 5,700 years, half of that remaining half will be gone. In another 5,700 years, half of that remaining quarter will be gone. The number 5,700 is called the *half-life*.

How long until all the C-14 is gone? If we do the math, we get a number that is less than 100,000 years. In fact, most of it disappears after only 50,000 years. (If the entire earth was made of C-14, it would all have turned into nitrogen in less than a million years.) This means that C-14 dating can only be used on objects that are young enough to still contain measurable C-14.

So what happens if C-14 dating is used on dinosaur bones? If they are millions of years old, they should have zero C-14 in them. People who firmly believe that dinosaurs are millions of years old don't even bother doing C-14 testing on dinosaur bones. (In fact, Jack Horner was offered grant money to completely cover all the costs of testing his *T. rex* bones for C-14. He turned it down. He claimed that C-14 testing was a waste of time because they already knew the age of the bones [66-69 million years]. To hear this conversation see the web address in the bibliography.) However, people who aren't so sure about the millions of years have gone ahead with C-14 testing. The results? No dinosaur bone, to date, has ever tested older than about 40,000 years. Many test at 20-30,000. Critics of this dating method say that the samples had been contaminated by "modern" carbon. However, the scientists doing the research had hired professional labs with excellent reputations. They were used to handling samples of very old objects, even items known to be several thousand years old. They said that they had ways to eliminate the possibility that contamination had affected the results.

After seeing these C-14 results, those questioning the millions of years took another look at radiocarbon dating. They tested zircons (those crystals that were used for uranium-lead dating) and found that they contained more helium atoms than they should. Why would rocks have any

TOO MUCH HELIUM?

helium at all? When uranium decays, it ejects a packet of 2 protons and 2 neutrons, called an *alpha particle*. If you add 2 electrons (easily found in any environment) to the alpha particle, you get a helium atom. Several of the decays on the way down to lead also eject helium nuclei. So it is expected, oddly enough, to find helium inside a zircon crystal.

Helium is a tiny atom (only hydrogen is smaller) so it is a slippery little thing that can manage to escape from a hard crystal. It is possible to predict the rate at which this leakage should happen. We can then predict how much helium should still be inside a zircon if it was millions or billions of years old. When actual tests were done on zircons to determine the amount of helium still inside them, the result was not even close to the prediction. The test result suggested that the zircons were less than 10,000 years old! So which should we use as our standard—the amount of lead or the amount of helium?

Let's take a break from physics and turn back to biology. You'll remember that we discussed how geology and biology are linked together because of fossils. Ever since the time of Darwin it has been assumed that the fossils must be showing us how one organism evolved into another. Dinosaurs were just one stop along the way.

The theory of evolution either works or it doesn't. It can't work just for the dinosaurs and not for everything else. When did evolution start? Most textbooks will tell you that the story of life started billions of years ago when the Big Bang happened.

First there was nothing, then suddenly, with a Bang, matter came into being. Where did matter and energy come from when there was nothing? Physicists have not worked that out yet. They just trust that someone in the future will figure it out, and then they go on with the story. So after all the quarks had formed atoms of hydrogen and helium, these elements condensed in certain areas to become stars. (There are serious problems with this, but beyond the scope of this book.) Inside stars, smaller elements turned into larger elements (more problems, but too

technical) and then stars exploded and elements were scattered across the universe. Some of these atoms ended up forming our sun and the planets, including the earth. (More problems, but too technical for this book.)

The earth was a hot, liquid ball at first, and took over a billion years to cool. Then (skipping a bit) it had cooled off enough for liquid water to exist on the surface and not boil off. Where the water came from no one is sure. Astronomers claim comets could have brought water, but really— have you seen how much water is in the oceans? Comets are tiny compared to the earth. (How about the other way around? Maybe water-bearing comets came from the water-covered earth?)

AI image of a molten earth

Evolutionists then imagine warm puddles on the surface of the earth where molecules could float around. Then, a lightning bolt hit one of the puddles and caused some of the molecules to attach themselves to other molecules. If this happened again and again, maybe larger and larger molecules would form.

Reality check—water isn't called the "universal solvent" for nothing. **Water dissolves and destroys molecules, especially organic molecules like proteins.** That's why people put dishes in the sink to soak overnight. Chemists who are doing "origin of life" research don't tell you that they have to pull their molecules out of their solutions quickly, before water tears them apart.

Just by random chance, amino acids hooked together to form proteins (except that special protein "tools" are need to do the attaching) and small proteins hooked together to make longer proteins (again, specialized protein "tools" are needed), and then, things like RNA, which has encoded information, came into existence. RNA is so improbable that many evolutionists look to astronomers again for help, hoping that maybe the beginnings of life came from outer space. This idea is called *panspermia*. At some point, RNA had to turn into DNA. This diagram might make it look simple, but this jump is far from simple. (Too technical to get into here.)

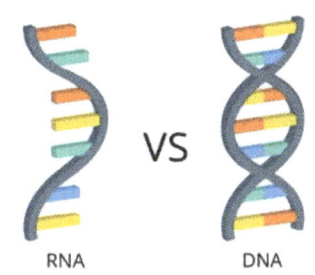

RNA
Ribonucleic Acid

VS

DNA
Deoxyribonucleic Acid

Scientists imagine that perhaps the early earth was a world inhabited only by RNA and simple proteins. Then, somehow, the RNA turned into DNA. However, this idea overlooks a critically important factor: information. Information is non-material, meaning it is independent of matter (atoms) and energy. Scientists who study the topic of information tell us that information automatically implies a sender and a receiver. (Can you think of any information in any form that did not have a sender?)

Those rungs on the DNA ladder (and on RNA) aren't random; they have a 4-letter code. Hundreds, thousands, or millions of rungs all arranged in a particular order. If that order changes, bad things happen. Also, DNA is useless if it is not in a complete cell that is surrounded by a protective membrane. To say that cells are complicated is a drastic understatement. The more we study cells, the more we find that even the simplest cells, such as bacteria, are more complicated than we ever imagined. Could a simple cell come into being just through random, undirected processes?

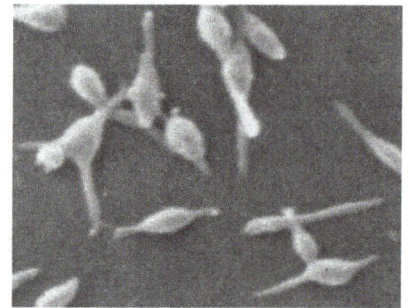

These mycoplasma bacteria used to be thought of as "simple" cells. We are now finding they aren't so simple.

This question has been evaluated from a purely mathematical standpoint. What are the chances of just one of a bacteria's proteins forming by chance? The amino acids in the protein have to be in just the right order. A scientist at MIT calculated this probability and came up with this answer: one chance in... not a million, not a billion, not a trillion, but a number so high that it doesn't have a name. It is a 1 followed by 65 zeros. This number is beyond the established definition by mathematicians of "impossible." The chances of three unrelated proteins forming all at the same time so that they could become part of a bacteria is three times that number: a 1 followed by 195 zeros. That's impossible times three.

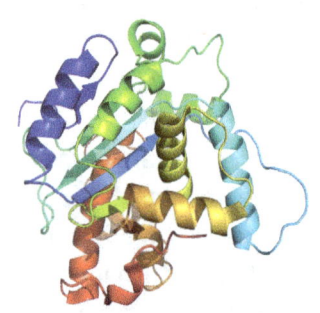

Proteins are often drawn like this. You can't see the amino acids.

44

We still don't have cells, let alone dinosaurs. Even if we just ignore how impossible it is to evolve a cell and go on with the story, there are still problems. The fossil record provides many of them. One of the bottom layers is called the **Cambrian** layer. Its name means "from the country of Wales." Many of the layers are named after places in Great Britain, where the early geologists lived. In the Cambrian layer we find all kinds of animals that lived at the bottom of seas or oceans. There are so many different types of animals that this layer is sometimes called the **Cambrian Explosion**. The fossil record goes from almost no life right to a layer with a wide variety. No transitional forms in between. If these animals all evolved from simpler animals, why is there no record of those simple ones?

This illustration shows an imagined scene from the Cambrian, based on fossils we have in that layer.

As we go up the geological column we find many similar gaps. There are complex animals with no record of all the many transitional forms that must have led up to them. Evolutionists like to point to particular examples that they think are well documented such as whale evolution. However, it has been revealed that several of the claimed transitional fossils are only partial fossils, with big pieces missing. In one case, the paleontologist who discovered the fossil took the liberty of adding a blowhole to the missing part of the skull so that it would look like a whale.

When it comes to dinosaur layers, such as Triassic, Jurassic and Cretaceous, there is more variety in the types of fossils than museums are telling their visitors. Generations of kids have grown up seeing paintings of dinosaurs wandering around amidst volcanoes and strange plants. However, as they dig out more and more fossils from these layers, many surprising fossils are turning up—both plants and animals that were thought not to have evolved yet during these dinosaur eras. For example, they have found fossilized ducks, flamingos, loons, parrots, sandpipers, frogs, squirrels, platypus, bees, cockroaches, and hundreds of small mammals that resemble animals such as rats, beavers and badgers. Grass has been found in fossilized dinosaur dung, disproving the idea that grass evolved millions of years after the dinosaurs. Pine trees and magnolia leaves have been found, and they, too, were thought not to have existed with dinosaurs. So far, no large mammals have been found, but they might not have lived in the same areas of the world as the dinosaurs. If we fossilized the world today, you would never find an elephant in the same place as a penguin or an iguana.

Another problem for evolution is finding genetic evidence for the "tree of life" idea. This idea started with a simple sketch from Darwin's notebook. Since then, the idea has been refined, with the tree becoming quite complex. The "trunk" of the tree represents the first simple organisms. The major "branches" represent different types of (now extinct) organisms that evolved from the first ones (plants or animals). The smaller branches represent animals or plants that evolved from the ones on the major branches. The plants and animals we have today are at the very tips of the tiniest branches. Scientists expected

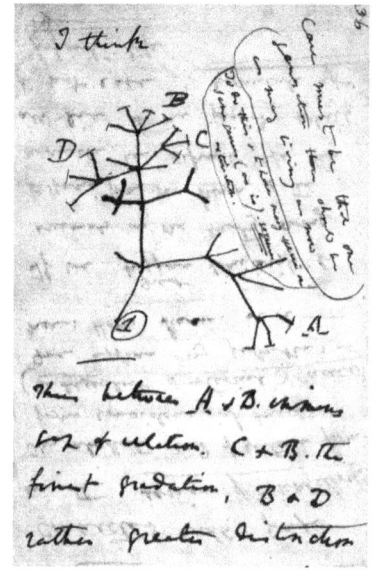

that as genetic evidence became available, it would prove this idea to be correct. But as we learn more about DNA, this has not been the case.

Geneticists have discovered DNA similarities in organisms that are on different branches of their tree. A simple example is echolocation. Both bats and dolphins can navigate using sound frequencies beyond the ability of any ear to hear. Even they can't hear them with their ears. They have special organs that send out and detect these high frequencies. Since DNA directs all the ways that cells and organs develop and function it makes sense that bats and dolphins would both have DNA for echolocation. However, this does not match up with the tree of life scheme.

Another simple example is striping patterns. Both the zebra and the zebra fish have black and white stripes. They both have DNA coding for this skin pattern. But zebras did not evolve from fish. Evolutionists get around this problem by making up a new theory called ***convergent evolution***. This is a fancy way of saying "it happened twice." Even though they admit that the probability of developing these traits even once is amazing, their unshakable belief in evolution demands that they propose that traits can evolve not just once, but many times.

What about the link between radiometric dating and biology? Is it ever wrong? There are some famous cases where radiometric dates were adjusted by millions of years when the biological evidence was too strong to ignore. In one case, they had found what looked like bird tracks in a rock layer in Argentina that had been radiometrically dated to 212 million years old. Because of the radiometric date, paleontologists said the tracks

had to be small dinosaur prints. This story was published in the most prestigious science journals in 2002. Then, about 20 years later, a new set of scientists examined the footprints and determined that they looked identical to the tracks of sandpipers (shown here).

The scientists had these rock layers re-dated using uranium-lead dating instead of the Ar-Ar method which had been originally used. The new radiometric date came back as 37 million years, well within the range for the proposed evolution of modern birds. These new dates were accepted, and a new articled was published with the younger results.

A second example involves a **hominid** (human or ape) fossil. A volcanic rock layer in eastern Africa was radiometrically dated to be 212-230 million years old. Then, the famous paleontologist, Richard Leakey, discovered a hominid fossil under that layer. This fossil was named specimen KNM–ER 1470. Evolutionists believe that hominids evolved less than 10 million years ago. Leakey believed that layer to be only 2-5 million years old. He requested more radiometric testing on the rock layers above the fossils. This

silly AI image of a hominid under volcanic rock

time the radiometric date came out as 2.6 million years. So Leakey dated his fossil at 2.9 million years. Then, a number of years later, the rock was again subjected to radiometric dating and this time it was only 1.9 million years old. Everyone accepted this date, even though it was more than 200 million years less than the orginal radiometric date.

Finally, if rock layers are not showing us the progress of evolution, then what might they be showing us? The order of the fossils, from bottom to top, lines up, in a general way, with levels of ecosystems, starting with the sea bottom then moving up through more shallow water, then marshes and seashores, and finally up onto dry land. Birds and large animals would be at, or close to, the top. This progression could explain the Cambrian Explosion. The fossils found in the Cambrian layer tend to be animals that live on the seafloor. Underneath these bottom-dwelling communities there's nothing but microscopic organisms—exactly what we find in the Precambrian layer. We are still left with the mystery of clams buried with dinosaurs and fossil graveyards containing both land and sea animals. These, and other mysteries, will be discussed soon.

CHAPTER 8: Many final thoughts

Scientists who favor the idea of a not-so-old earth see dinosaur soft tissue as powerful evidence that dinosaurs aren't as old as we've been led to believe. Forensic science seems to support the idea that organic molecules could never last even a million years under ideal conditions, let alone the harsh enviroments where the bones are found. Scientists who favor the old-earth-and-evolution idea point to radiometric dating as the final answer. An impartial spectator would say that the young earth scientists will need to come up with an alternative explanation for the rock layers and the old earth scientists will need to find a way that stretchy collagen can last for millions of years.

Some of the old earth advocates have refocused their research from simply discovering more biomaterial to trying to figure out a way that these structures might possibly be preserved for millions of years. We already mentioned the proposal that iron, presumably from hemoglobin in blood, was involved in the preservation. We also saw that this idea really didn't solve the problem. Many tissues were not close to blood cells, the amount of iron from the blood was not enough, and experiments

Ostrich tissues were used in iron experiments.

using iron and modern bird tissue were not convincing. Additionally, Mark Armitage's work showed clotting of blood in the vessels, suggesting that the blood cells and their contents did not leak out into the bone tissue.

The most recent idea is that collagen's amino acid structure is so tight (remember those braided strands on page 5) that water, the "universal solvent" that dissolves just about anything, is not able to get into the spaces between the atoms to dissolve the collagen in the bones. The most likely place that water would be able to tear apart a protein is the place where one amino acid connects to another. On the next page you will see a diagram of how two amino acids are joined. This diagram will help you understand what these researchers are proposing.

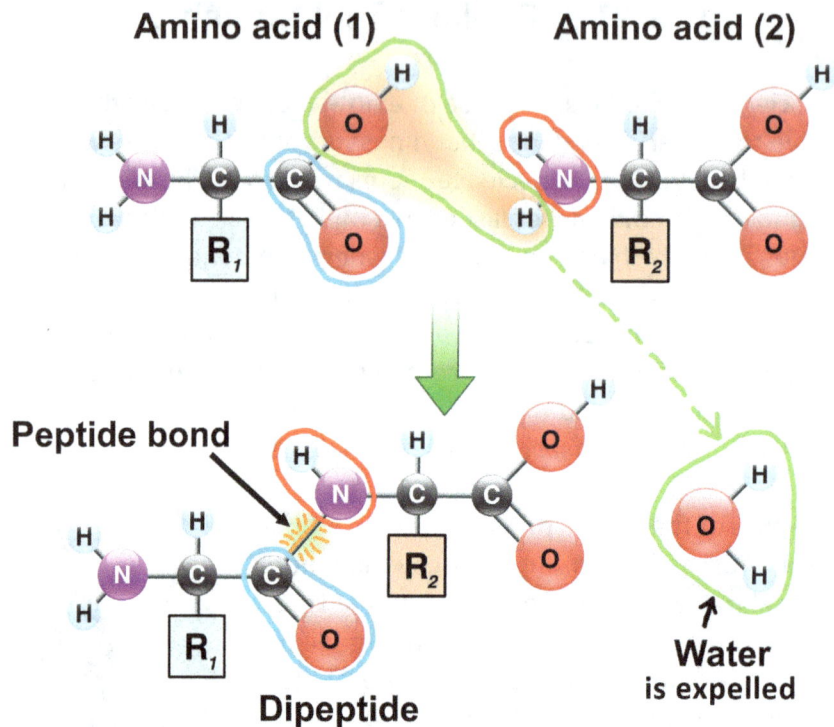

Amino acid (1)   **Amino acid (2)**

Peptide bond

R_1

R_2

Dipeptide

Water is expelled

On the top you see two separate amino acids. The R is the same thing as the ? question mark in the previous amino diagrams. A green circle has been drawn around the atoms (one oxygen and two hydrogens) that will be removed from the molecules as they join together. The dashed green line shows that they end up forming a water molecule that will just float away. The blue and red lines are circling groups of atoms to keep your eye on. The green arrow indicates progression from apart to joined. In the bottom half you see that the two aminos are now connected, as the carbon atom from amino (1) attaches to the nitrogen atom in amino (2). The bond between them is called a ***peptide bond***.

This process can be reversed. You can go from the bottom to the top. A special little protein tool that acts like a scissor can cut the peptide bond and then chop a water molecule in half and use the H and OH to patch the ends that were cut, returning it to the formation seen in the top half of the diagram (two separate aminos).

Bonding between atoms is done through their electrons, although they are not shown in the diagrams on the previous page. Below is a diagram showing just a few of the electrons in this molecule. The important electrons for this discussion are the ones that the arrow is pointing to.

electrons

That oxygen's electons are the key. As we saw on page 5, collagen is made of things that look like bundles of braided ropes. This peptide molecule shown above will be one of thousands, in a long line, and it will be twisted together with two other long lines. The unpaired electrons (of the O) will sit very tightly next to a neighboring strand. Those electrons can interact with the neighbor strand, forming a kind of "glue" to keep the braid together. They say that this "gluing" by the electrons can keep water away from the peptide bonds, and therefore allow it to survive for a very long time.

However, collagen is not the only molecule that has survived. DNA, RNA, and pigment molecules don't have the collagen advantage, and neither did that bacteria from a termite's gut that came back to life. Additionally, there is another factor to be considered: natural radiation. We are un-

aware of all the radioactivity around us, coming from rocks under our feet as well as some rays from outerspace. Short term, this doesn't affect us very much. But over millions of years, the radiation would degrade all organic molecules, including DNA, RNA, pigments and collagen.

Those who favor ages of only thousands of years for the dinosaurs point to other questions that arise when considering all the factors that go into fossilizataion. Why did so many animals get fossilzed? There aren't any fossils forming today. Fossilzation takes special conditions. You need rapid burial so that the organism doesn't rot. (There are fossilzed jellyfish—how long do those last on the beach?) You need the burial to be with sediment that is full of water saturated with minerals such as calcium carbonate, which hardens into limestone, and silicon dioxide, which hardens into sandstore. The water found in today's rivers, lakes and oceans, doesn't have a high enough amount of these minerals to fossilize anything. We can watch whale carcasses rotting on the ocean floor (and a group of scientists actually did this!) and we know they don't fossilize. Yet there are thousands upon thousands of fossilized fish, sharks, rays, whales, and many extinct sea creatures. Even entire schools of fish can be found in the fossil record, looking as though they were buried instantly, as they were swimming along.

whale fossil in the Sahara

rotting whale carcass

Also, we have millions of fossilzed clams found in the closed position. If you've ever been to the beach, you know that living clams are tightly closed, and are just about impossible to force open. A strong muscle keeps the two halves together. When the clam dies, that muscle also dies, and the shell falls open. Not long after, the connective tissue at the joint dissolves and the two halves get scattered across the beach. Therefore, fossilized clams with closed shells must have been buried while they were still alive.

Speaking of clams, did you know that there are millions of closed-shell fossilized clams at Dinosaur National Monument? How did living clams from the ocean get buried with these air-breathing giant dinosaurs? There are also clam shells on top of Mt. Everest. How did they get there?

Another interesting factor to consider is the position in which many dinosaur skeletons are found, with their necks arched backward. This is so common that it has become known as the "dead dino posture." This position is called **opisthotonus** and is seen in both animals and humans in certain health conditions such as tetanus, epilepsy, cerebral palsy, poisoning and drowning. All of these conditions can cause the muscles in the spine to tighten so much that the entire spine curves backward. To confirm that this can happen in the case of drowning, a scientist decided to immerse recently dead chickens in cold water. As expected, all of them assumed the dead dino posture.

So did these dinosaurs all have tetanus or epiletic seizures? Were they poisoned? Or did they die due to drowning? The simplest and most obvious answer is that they drowned.

Mark Armitage's work with blood clotting in fossilized dinosaur blood vessels also points to drowning. The type of clotting he is seeing is called **disseminated intravascular coagulation** and is known to occur in human drowning victims. Mark finds these clots in bones from different layers, from Permian to Cretaceous, spanning an assumed time period of over 150 million years. It doesn't matter what layer he examines, he

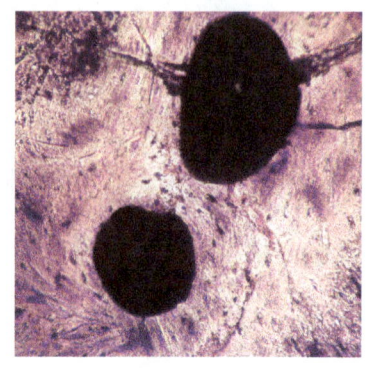

finds the same blood clots, the same evidence of drowning.

Another strong piece of evidence to consider is the previously mentioned stench of Hell Creek. How can animals that have been dead for millions of years still stink? Stinky smells are made by bacteria. What are the bacteria eating? They can't live on nothing. There must be organic molecules still inside the bones. Over the course of 65 million years or more, surely the bacteria would have finished off all the available food, then died. Bacteria aren't immortal.

What about all the layers of rock? Could they have formed quickly? The full answer to this would take several more pages to explain, but the short answer is: "Yes." The research into this question has been done mostly by scientists who already favor a younger age for rock layers because those who favor millions of years don't see any point in doing experiments about something that they already believe is false. Researchers use a device called a **flume** in order to determine whether layers of sediment could form all at the same time, not one at a time. The flume is long tank in which water flows in one direction. Many types of sediment can be added, and then the researchers can watch to see what happens. They saw particles sort themselves out, with similar particles grouping together and all falling out of the water at the same time, creating striped layers. In one experiment, the researcher saw thin layers forming one on top of the other within minutes. This challenges the idea that it takes millions of year to form rock layers. (Articles about this are listed in the bibliography.)

What about radiometric dating? As was mentioned earlier, all methods of radiometric dating involve making assumptions about what the conditions were a long time ago, when there weren't any scientists there to observe and take notes. We must assume that conditions were the same then as they are now. But were they?

We know for sure that volcanic rocks often have too much Argon-40 in them because we've tested rocks from recent volcanc eruptions. For example, Mt. Hualalai in Hawaii erupted in the year 1800 and its rocks tested at about 1.5 million years old. Mt. Etna erupted in 1792 and radiometric dating said 350,000 years old. Mount St. Helens erupted in 1980 and gives dates ranging from about 100,000 years to 1.5 million. Scientists concluded that Ar-40 from volcanic gases must have gotten mixed into the lava during the eruption, putting extra Ar-40 into the rocks. (Remember, more Ar-40 will result in an older date.) Also, a diamond from Zaire was dated at 6 billion years old using the K-Ar dating method. This is 1.5 billion years older than the assumed age of the earth! Obviously, there was extra Ar-40 in that diamond. Since we know that all these rocks must contain extra Ar-40, why do we assume that all the Ar-40 in what we think is a very "old" rock came only from radioactive decay over time?

In the Grand Canyon there is a location where there is volcanic rock at the base of the canyon underneath many rock layers. Nearby, there is volcanic rock from a lava flow that flowed from the surface downward across all those rock layers. The flow from the surface must be much younger than the rock layers underneath it. Yet, radiometric dating gives about the same ages for both volcanic layers, top and bottom. The answer? The radiometric data is showing us that these rocks came from the same lava source. The amounts of radioactive decay products have resulted from the chemistry of the magma that produced the rocks, instead of being the result of time.

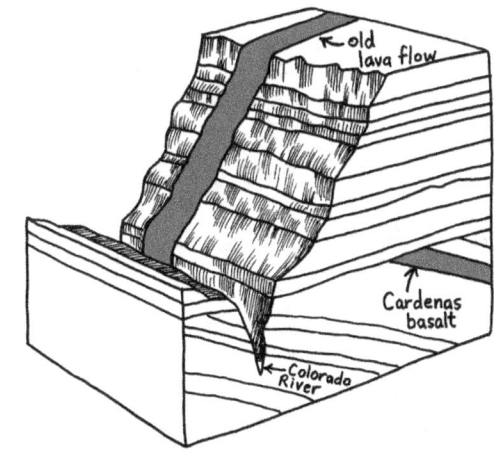

So back and forth we can go, with arguments and counter arguments from each side. Sometimes, a person is won over to one side or the other by just the scientific data. However, most of the time, the deciding factor will be arguments from other places—from personal beliefs about how the earth and the universe came into being. We've already outlined the popular belief in the Big Bang as the source of everything. We've discussed the theory of evolution as one attempt to explain the origin of all living things. Do any scientists believe otherwise?

The "fathers " of many branches of science didn't believe in either the Big Bang or evolution. In some cases, these theories didn't exist yet when that scientist was alive, but in other cases, these ideas were already floating around. Here is a list of scientists with their personal beliefs listed in the center column. The main point here is that belief in evolution and the Big Bang is not necessary in order to be a good scientist.

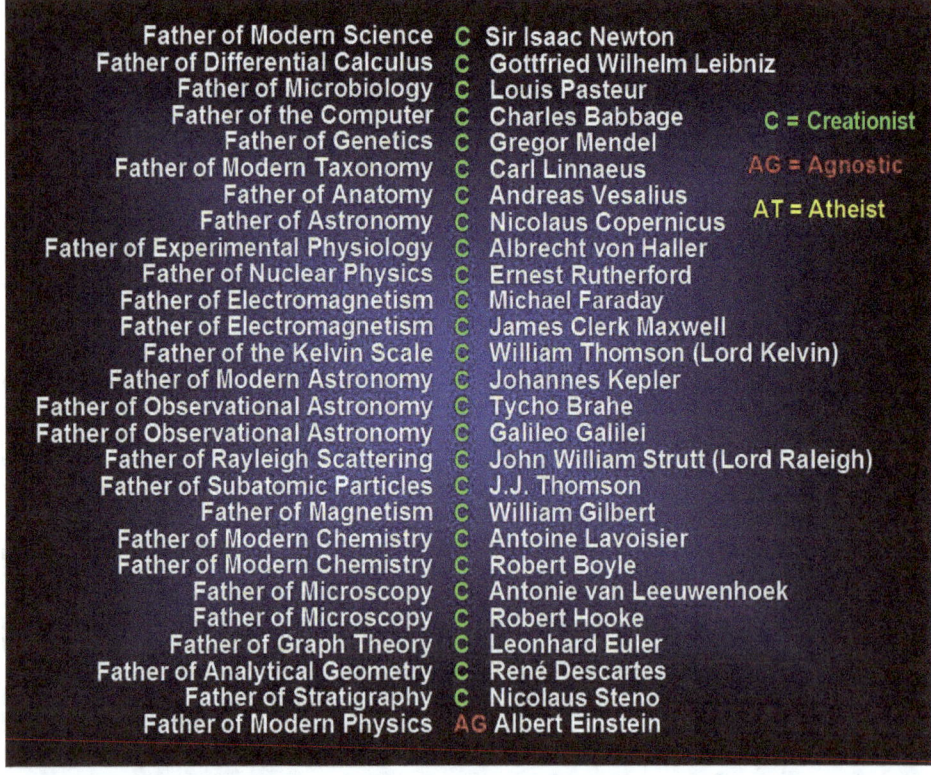

Father of Modern Science	C	Sir Isaac Newton
Father of Differential Calculus	C	Gottfried Wilhelm Leibniz
Father of Microbiology	C	Louis Pasteur
Father of the Computer	C	Charles Babbage	C = Creationist
Father of Genetics	C	Gregor Mendel
Father of Modern Taxonomy	C	Carl Linnaeus	AG = Agnostic
Father of Anatomy	C	Andreas Vesalius
Father of Astronomy	C	Nicolaus Copernicus	AT = Atheist
Father of Experimental Physiology	C	Albrecht von Haller
Father of Nuclear Physics	C	Ernest Rutherford
Father of Electromagnetism	C	Michael Faraday
Father of Electromagnetism	C	James Clerk Maxwell
Father of the Kelvin Scale	C	William Thomson (Lord Kelvin)
Father of Modern Astronomy	C	Johannes Kepler
Father of Observational Astronomy	C	Tycho Brahe
Father of Observational Astronomy	C	Galileo Galilei
Father of Rayleigh Scattering	C	John William Strutt (Lord Raleigh)
Father of Subatomic Particles	C	J.J. Thomson
Father of Magnetism	C	William Gilbert
Father of Modern Chemistry	C	Antoine Lavoisier
Father of Modern Chemistry	C	Robert Boyle
Father of Microscopy	C	Antonie van Leeuwenhoek
Father of Microscopy	C	Robert Hooke
Father of Graph Theory	C	Leonhard Euler
Father of Analytical Geometry	C	René Descartes
Father of Stratigraphy	C	Nicolaus Steno
Father of Modern Physics	AG	Albert Einstein

This list cuts off at about the turn of the 20th century. Since then there have been more scientists who identify themselves as atheists, believing that the universe and everything in it is a result of nothing more than blind chance and random processes. Since the mid-1900s, the Big Bang and evolution have been taught as fact in schools, so the majority of scientists at any university today grew up being taught the atheistic world view. Thus, it's not surprising that the majority view right now is that dinosaurs lived millions of years ago and evolved from lower forms of life. This view has been pushed so hard by schools, movies, television, and museums that even faith-based groups have taken up this view. Many religious denominations have been convinced that radiometric dating is the ultimate source of truth about the history of the world. But what if radiometric dating isn't showing us time, but something else? What if it is showing us information about the chemistry of the earth, or the chemistry of lava flows? Are today's scientists able to entertain alternative ideas?

There are many examples in the history of science when scientists truly believed something that was later proved to be false. Until the late 1700s, scientists believed in a fictional substance called "phlogiston" that was said to be the cause of fire. Finally, Antoine Lavoisier (one of the scientists listed in the chart) gave such convincing proof of the real nature of fire that the phlogiston theory had to be abandoned. For several hundred years, doctors believed that the way to cure someone was to make them bleed. George Washington died because of this false medical belief. Doctors also believed that good hygiene was not necessary, and even after there was convincing proof that hand washing saved lives, many doctors still refused to wash their hands. Geophysicists laughed at the man who suggested that the continents had drifted apart. They publicly ridiculed him for decades before they decided...oops, he was right. Geologists spent a fair portion of the 20th century rejecting a theory that is now accepted as the **Missoula Flood**. Science is supposed to test new ideas, not persecute them. Shouldn't we ask, "How do I arrive at the truth? Do I have all the evidence or is something missing? What is new evidence suggesting?" We'd like to believe that science and scientists are completely objective, but history has shown us that this isn't true.

It's interesting to stop and note that everyone believes that something came from nothing. We all agree on that. The Big Bang says that something came from nothing. Creationism says that something came from nothing. We don't have to argue that point. **Science can't prove or disprove any ideas about how the world and the universe came into being**. These events happened before humans existed. We can't do experiments on the past, only the present. The present laws of physics, as we observe them, say that matter (stuff made of atoms) and energy cannot be created or destroyed, they can only change from one form to another. **It's called the Law of Conservation of Matter and Energy.** So if physics says that something can't come from nothing, then how did it happen?

Creationism's answer to the "how" question is that an intelligent Creator was able to plan how quarks would form atoms, how atoms would work together to form molecules, how molecules would be able to fit

together to make proteins, how amino acids would line up in just the right order to form complex proteins, how proteins would work together to form cells, how a great variety of cells would work together to form tissues and organs, how all the organs would function as one unit in a whole body, and how bodies of different kinds (plants, animals and humans) would form complex ecosystems. Creationism doesn't need eons of time to explain the dinosaurs.

Creationism says that dinosaurs were made to be majestic creatures that displayed the Creator's power and wisdom. In fact, dinosaurs are in the Hewbrew scriptures. In the book of **Job** *(Jobe)*, there is a description of an animal that can't possibly be anything we see living today. In the 40th chapter of this book, we read that God reminds Job how wonderful creation is. He talks about animals that Job was familiar with, such as the lion, the raven, the mountain goat, the donkey, the wild ox, the ostrich, the horse, the hawk and the eagle. Then He says:

"Look at the 'Behemoth' which I made in the same way I made you *(meaning humans)*. He eats grass like an ox. Look at the strength in his loins, and the power in the muscles of his belly! He makes his tail stiff like a cedar tree; the sinews *(tendons and ligaments)* of his thighs are closely knitted together. His bones are like tubes of bronze, his limbs *(legs)* are like bars of iron. It is a prime example of the work of God, and only its Creator is able to threaten it. The hills offer it their best food, and all the wild animals play nearby. Under the lotus plants it lies, hidden among the reeds in the marsh. The lotus plants give it shade among the willows by the stream. It is not disturbed by the raging river, and it is not concerned if the Jordan river rushes against its mouth. Can anyone capture it by the eyes *(catch it off guard)* or trap it and put a ring in its nose to lead it away?"

(This rendition is a combination of several translations.)

What does this sound like to you? It can't be an elephant because it has a tail like a cedar tree. That also rules out a hippo. It can't be an alligator or crocodile because they don't eat grass like an ox. Could it be... a sauropod, such as an apatosaurus or a brachiosaurus? They definitely have tails like cedar trees and we are pretty sure they were plant eaters. With bodies so large, they likely spent time in water whenever they could. Sauropods fit the description better than any other animal. Job apparently had seen this animal and knew what it was. We know for sure that this story was written down several thousand years ago, as it appears in the famous Dead Sea Scrolls.

We have no reason to believe that a mythological animal is being described here. Behemoth is included in a list of animals that are not mythological. The description of Behemoth sounds like a realistic description; it tells us what the animal ate, what the body was like, and where it preferred to spend time (in marshes). If we take this text at face value, it implies that thousands of years ago, humans saw at least some dinosaurs. They might not have been as large as the huge fossils we find, but even if Job's sauropod was half the size of the large skeletons, it would still have been the largest land animal.

If you want to read about another dinosaur in the book of Job, you can do a quick Internet search with key words "Job chapter 41." Compare this description to animals like the mosasaurus. Ancient cultures told tales of terrifying sea monsters that were definitely not whales or sharks. Could their exaggerated stories have been based on real (much smaller) animals?

Another interesting observation involving a list of animals is the Asian lunar calendar. The calendar goes in 12 year cycles, with an animal for each year. The animals are: rat, tiger, ox, rabbit, dragon, snake, horse, goat, monkey, rooster, dog, pig. Why would the ancient Asian peoples put a mythological animal into their list of real animals?

Many cultures have legends about animals that fit the description of what we would call a dinosaur. After the real "dragons" went extinct (as many animals have, including the mammoth and the dodo bird) each retelling of the legend got more and more fanciful. Chinese folklore dragons don't look like real animals. Without a living "dragon" for comparison, there was no reality check on the stories.

So why mention these legendary stories? These can be one of the many pieces of evidence that someone might use in their investigation of how old dinosaur soft tissue might be. If even a few dinosaurs were still living during ancient times, this means that they didn't all die out millions of years ago. (Scientists scoff at this type of evidence, so be prepared. Any suggestion that challenges their assumptions will be called "ridiculous.")

In the book of Job, the Bible not only mentions dinosaurs, it also tackles one of life's hardest questions—the very thorny question of why bad things happen to good people. This is a question we still ask, thousands of years later. In other books, the Bible addresses questions like: "What is my purpose in life?" "What happens when I die?" "Why is there evil in the world?" Because the Bible gives solid answers to these hard questions, many people are willing to trust it for answers to questions like "How old is the earth?" and "Did we evolve?"

For those who believe in a young earth, it's all one big package—the same book that gives answers to philosophical and moral issues also says that we were created and gives a timeline from the first people, Adam and Eve, up through historical people that we can firmly date, such as Daniel, Jesus, and St. Paul. You can't chop up the Bible, which is a continuous story from start to finish, and discard parts you don't like. They point out that people who claim to believe the Bible and yet still try to hold on to millions of years and evolution are rejecting what the Bible says very plainly about when death came into the world. In the beginning, everything was good. Death came into existence after sin and evil entered the world through the first two humans. Evolution makes death into a good and necessary thing that drives the whole evolutionary process with unfit animals dying off (even though we can't prove that any particular fossilized animal was "unfit"). However, we all sense instinctively that death is not good. Think about all those poor dinosaurs who died in agony with arched backs. Death is ugly. If you add to all the sadness and ugliness of death the mathematical improbability of evolution and the witness of forensic science, creationism doesn't seem so ridiculous. Also, creationism as presented in the Bible actually offers a remedy for death—hope for a dying world. If you are interested in the details, you might start with searching: "Genesis 1" and then "book of John."

SUMMARY

Original biomaterial in fossils was first discovered in the 1950s. Discoveries continued to be made through the following decades, but this topic was not brought to the attention of the general public until Mary Schweitzer made her discoveries in 2004. She found several types of "soft tissue" in a *T. rex* bone, including collagen that was still stretchy. After 2004, this field of research took off and today many research papers are published every year. To date, the following biological materials have been discovered: amino acids, small proteins, pigment molecules, several types of collagen, osteocytes, nerve fibers, blood cells, plus indirect evidence for the presence of heme, hemoglobin, DNA and RNA. Dormant bacteria were also found in the gut region of a termite fossilized in amber (tree sap that is now hard as rock).

These exciting finds have led to a considerable amount of controversy. At first, Mary was sharply criticized by mainstream scientists who didn't think biological material could survive for millions of years. Mary found herself at the center of a swirl of controversy in her own scientific community. Mark Armitage was also a victim of the soft tissue controversy when he was fired after publishing his tissue findings in a scientific journal. Not long after, a new level of controversy opened up when creationists restated the claim that these tissues could not survive for millions of years, and were proof that dinosaur fossils were only thousands of years old.

Evolutionists and creationists have different views on fossil formation. Evolutionists believe that fossilization takes place over long periods of time. Creationists believe that fossil formation was the result of a global flooding event that destroyed the surface of the earth and buried many life forms in sediments saturated with water and minerals, a perfect recipe for fossilization.

Old earth (evolution) advocates believe in radiometric dating as the foundation of their worldview, and they trust that at some point in the future, the mystery of the origin of life will be solved.

Young earth (creation) advocates have Creation as the foundation of their worldview, and they trust that at some point in the future the mystery of radiometric dating will be completely solved.

BIBLIOGRAPHY

Website I consulted most often for a good list of papers on fossilized biomaterial:

https://kgov.com/dinosaur-soft-tissue-original-biological-material
Direct link to this list on google docs: https://docs.google.com/spreadsheets/d/
1BSM-oQJXxhYBlsLE3gGl3bz8GXgtoLy-oLOsSNF_Lhw/edit?pli=1&gid=0#gid=0

Papers about biomaterial finds that I gave special consideration in this book:

1954
Abelson. Carnegie Institution of Washington, D. C. Year book No. 53
(July 1, 1953-June 30, 1954)

1962
Isaccs, W., Little, K., Cureey, J. et al. Collagen and a Cellulose-like Substance in Fossil
Dentine and Bone. Nature 197, 192 (1963). https://doi.org/10.1038/197192a0

1966
Pawlicki, R., Korbel, A. & Kubiak, H. Cells, Collagen Fibrils and Vessels in Dinosaur Bone.
Nature 211, 655–657 (1966). https://doi.org/10.1038/211655a0

1987
Davies, Kyle L. Duck-Bill Dinosaurs (Hadrosauridae, Ornithischia) from the North Slope of
Alaska. Journal of Paleontology, Vol. 61, No. 1 (Jan., 1987), pp. 198-200.

1991
Gurley, L.R., Valdez, J.G., Spall, W.D., Smith, B.F., & Gillette, D.D. Proteins in the fossil
bone of the dinosaur, Seismosaurus. J Protein Chem. 1991 Feb;10 (1):75-90.
doi: 10.1007/BF01024658. PMID: 2054066.

1992
DeSalle R, Gatesy J, Wheeler W, & Grimaldi D. DNA sequences from a fossil termite
in Oligo-Miocene amber and their phylogenetic implications. Science. 1992 Sep
25;257(5078):1933-6. doi: 10.1126/science.1411508. PMID: 1411508.

1994
Berchtold M., Ludwig W., & König H. 16S rDNA sequence and phylogenetic position of
an uncultivated spirochete from the hindgut of the termite Mastotermes darwiniensis
Froggatt. FEMS Microbiol Lett. 1994 Nov 1;123(3):269-73.
doi: 10.1111/j.1574-6968.1994.tb07235.x. PMID: 7527363.

1995
Cano, Raúl J. & Borucki, Monica K. Revival and Identification of Bacterial Spores in 25 to 40-Million-Year-Old Dominican Amber. Science, 19 May 1995. Vol 268, Issue 5213.

1997
Schweitzer, Mary Higby, et al. Preservation of Biomolecules in Cancellous Bone of Tyrannosaurus Rex. Journal of Vertebrate Paleontology, Vol. 17, No. 2, 1997, pp. 349–59. JSTOR, http://www.jstor.org/stable/4523811. Accessed 26 Jan. 2026.

2005
Schweitzer, Mary H., et al. Soft-Tissue Vessels and Cellular Preservation in Tyrannosaurus rex. Science 307, 1952-1955 (2005). DOI:10.1126/science.11083

2007
Soft tissue in 10 mya frog bones: https://pubs.geoscienceworld.org/gsa/geology/article-abstract/34/8/641/129601/High-fidelity-organic-preservation-of-bone-marrow?redirectedFrom=fulltext

2013
Schweitzer M.H., Zheng W., Cleland T.P., & Bern M. Molecular analyses of dinosaur osteocytes support the presence of endogenous molecules. Bone. 2013 Jan;52(1):414-23. doi: 10.1016/j.bone.2012.10.010. Epub 2012 Oct 17. PMID: 23085295.

2016
Armitage, Mark. Preservation of Triceratops horridus Tissue Cells from the Hell Creek Formation, MT. Microscopy Today, Jan. 2016. doi: 10.1017/S15519295515001133

2020
Bailleul, Alida M., Zheng, Wenxia, Horner, John R., Hall, Brian K., Holliday, Casey M., & Schweitzer, Mary H. Evidence of proteins, chromosomes and chemical markers of DNA in exceptionally preserved dinosaur cartilage. National Science Review, Volume 7, Issue 4, April 2020, Pages 815–822, https://doi.org/10.1093/nsr/nwz206

2020
Armitage, Mark. UV Autofluorescience Microscopy of Dinosaur Bone Reveals Encapsulation of Blood Clots within Vessel Canals. Microscopy Today, September 2020 doi: 10.1017/S1551929520001340

2023
Armtiage, Mark. First Report of Peripheral Nerves in Post-Cranial Elements of *Cacops* Williston, 1910, (Temnospondyli: Dissorophidae) from the Lower Permian Richards Spur, OK. Microscopy Today, January 2023.

Websites that I found helpful for these topics:

Meduallary bone: https://en.wikipedia.org/wiki/Medullary_bone

Article on Mary Schweitzer's discoveries:
   https://barryyeoman.com/2006/04/schweitzers-dangerous-discovery/

James Hutton: https://en.wikipedia.org/wiki/James_Hutton

Osteocytes https://en.wikipedia.org/wiki/Osteocyte

Dead Dino Posture: https://en.wikipedia.org/wiki/Opisthotonus

For a complete list of Mark Armitage's papers: https://dstri.org/articles-updates/

Radiometric dating:
https://en.wikipedia.org/wiki/Uranium%E2%80%93lead_dating
https://en.wikipedia.org/wiki/Zircon
https://en.wikipedia.org/wiki/K%E2%80%93Ar_dating
https://creation.com/en/articles/helium-evidence-for-a-young-world-continues-to-
   confound-critic (answers to objections about too much helium)
https://creation.com/en/articles/radiometric-backflip (sandpiper tracks)
Lecture at 2017 "Is Genesis History" conf. by Andrew Snelling:
   https://www.youtube.com/watch?v=5la7SoO6FfY

Details about C-14 dating:
https://creation.com/en/articles/c14-dinos (report censored at Geophys conference)

Abiogensis (origin of life):
https://kgov.com/abiogenesis-royal-truman
Analysis of problems with updated Miller-Urey experiment (Bada, 2008)
   https://creation.com/en/articles/miller-urey-revisited-oxidizing-atmosphere
Proteins forming by chance: https://creation.com/en/articles/divining-design#forwardref-23

Out-of-order fossils:
https://creation.com/en/articles/fossils-out-of-order
   (This webpage has links to secular research that backs up these claims.)
https://creation.com/en/articles/modern-birds-with-dinosaurs
Fossil mammal and dino together: https://www.nature.com/articles/s41598-023-37545-8

Preservation of collagen:

https://www.acs.org/pressroom/presspacs/2024/september/why-dinosaur-
collagen-might-have-staying-power.html
https://news.mit.edu/2024/mit-chemists-explain-why-dinosaur-collagen-survived-
millions-years-0904
https://www.siliconrepublic.com/innovation/how-did-collagen-survive-in-dinosaur-fossils
https://interestingengineering.com/science/mit-dinosaur-collagen-survival-secret

Fakes used to support whale evolution:

Phil Gingerich admits Rodhocetus lacks imagined whale features:
https://www.youtube.com/watch?v=POMjpTSk3no
Dr. Hans Thewissen interviewed about blowhole of Ambulocetus:
https://www.youtube.com/watch?v=ObX3UdrKvZo

Fast formation of sedimentary rock layers: (references at bottom of articles)

https://creation.com/en/articles/experiments-on-stratification-of-heterogeneous-sand-mixtures
https://creation.com/en/articles/sedimentation-experiments-is-extrapolation-
appropriate-a-reply
https://ianjuby.org/sedimentation/
https://eos.org/research-spotlights/specious-timescales-from-sedimentary-layers
https://www.icr.org/article/15415/

Website offering 10 million dollar prize for theory about origin of life: ev02.org

Books:

DeYoung, Don. *Thousands... Not Billions*. Master Books, 2005. ISBN 978-0-89051-441-2

Johnson, Donald E. *Programming of Life*. Big Mac Publishers, 2010. ISBN 9780982355466

Sibley, Andrew. *England's Jurassic Coast; Rocks, Fossils, and Biblical History*. Creation
Book Publishers, Powder Springs, GA. 2025 ISBN 978-1-954198-17-3

Audio resources:

Radio show interviewing Mark Armitate about getting fired from Cal State:
https://kgov.com/pji-brad-dacus-triceratops-tissue

Mark Armitage discussing bloot clots, vessels, and nerve tissue found in dinosaur bones:
https://kgov.com/dinosaur-blood-clots

Bob Enyart offers Jack Horner at 23K grant to carbon date dino bones:
https://www.youtube.com/watch?v=PXy7EH13lCo&t=16s